Tropical Cyclone Future Projections

Tropical Cyclone Future Projections

Editor

Hiroyuki Murakami

MDPI • Basel • Beijing • Wuhan • Barcelona • Belgrade • Manchester • Tokyo • Cluj • Tianjin

Editor
Hiroyuki Murakami
University Corporation for
Atmospheric Research (UCAR)
USA

Editorial Office
MDPI
St. Alban-Anlage 66
4052 Basel, Switzerland

This is a reprint of articles from the Special Issue published online in the open access journal *Oceans* (ISSN 2673-1924) (available at: https://www.mdpi.com/journal/oceans/special_issues/tropical_cyclone).

For citation purposes, cite each article independently as indicated on the article page online and as indicated below:

LastName, A.A.; LastName, B.B.; LastName, C.C. Article Title. *Journal Name* **Year**, *Volume Number*, Page Range.

ISBN 978-3-0365-3218-9 (Hbk)
ISBN 978-3-0365-3219-6 (PDF)

Contents

Article

Statistical Decomposition of the Recent Increase in the Intensity of Tropical Storms

Stephen Jewson [1],* and Nicholas Lewis [2]

[1] Risk Management Solutions Ltd., Peninsular House, 30 Monument Street, London EC3R 8NB, UK
[2] Bath, UK; nhlewis@btinternet.com
* Correspondence: Stephen.Jewson@gmail.com

Received: 21 October 2020; Accepted: 7 December 2020; Published: 11 December 2020

Abstract: In a recent paper, Kossin et al. showed that during the period from 1979 to 2017, there was a statistically significant increase in the ratio of category 3–5 to category 1–5 tropical storm fixes in the ADT-HURSAT satellite dataset of tropical cyclone observations. The sign of this increase is consistent with previously developed theory and modelling results for how tropical cyclones may change due to climate change. However, without further analysis, it is difficult to understand what the implications of this increase might be for present day tropical cyclone risk. It is also difficult to understand how tropical cyclone risk models might be adjusted to reflect this increase, since this ratio is not typically directly represented in such models. Our goal is therefore to understand the drivers for this increase in terms of changes in the numbers of fixes of different categories of storms in different basins, which are quantities that are more directly related to tropical cyclone risk and risk modelling. We use both heuristic and quantitative methods. We find that the increase in the ratio is mainly driven by a decrease in the denominator (the number of category 1–5 fixes) and to a small extent by a slight increase in the numerator (the number of category 3–5 fixes). The decrease in the denominator is mostly driven by a statistically significant reduction in the number of category 1 fixes outside the North Atlantic. The slight increase in the numerator is mostly driven by a statistically significant increase in the number of category 3–4 fixes in the North Atlantic. Based on these results, we discuss different ways in which the increase in the ratio could be represented in risk models.

Keywords: tropical cyclones; hurricanes; climate change; tropical cyclone fixes; trends; ADT-HURSAT

---

1. Introduction

The effects of climate change on temperature and sea-level are mostly well understood and readily quantified, albeit with uncertainty. The effects of climate change on other variables, however, are generally less well understood. In this article, we consider aspects of the possible effects of climate change on the frequency and intensity of tropical cyclones, with a particular focus on quantities that are directly of relevance to tropical cyclone risk and risk modelling. There are essentially three methodologies that one can use to investigate the question of how the frequency and intensity of tropical cyclones might change. One is theory. There is no complete theory of tropical cyclones that can give good explanations for all the aspects of tropical cyclone behavior that are seen in nature. However, for the intensity of tropical cyclones, there is a body of theory [1] that can be used to attempt to predict how tropical cyclones might change in the future: the implications of this theory are that the intensities of individual tropical cyclones may have already increased, and may continue to increase, all other things being equal [2]. The second methodology that one can use to understand the possible effects of climate change on tropical cyclones is dynamical modelling of the climate using climate models. In a recent survey of climate model results for tropical cyclones [3], different models were found to give somewhat different results, highlighting the uncertainty, but many models showed decreasing overall frequency of tropical cyclones, the highest resolution models suggested an increase in the frequency of very intense tropical cyclones (strength categories 4 and 5), and most models showed an increase in the proportion of tropical cyclones that reach very intense. The third methodology that one can use to understand how tropical cyclones may change under a changing climate is to look at the past, and use statistical analysis to assess whether it is already possible to see any changes in tropical cyclone behavior that could be due to climate change. This is difficult for three particular reasons: tropical cyclones, especially the most intense tropical cyclones, are rare; past measurements of tropical cyclone behavior are inhomogeneous in coverage and quality [4]; and tropical cyclone behavior is affected by climate variability on multiple time-scales, much of which is not fully understood, and which hinders the detection of long-term trends [5]. As a result, there is no clear consensus, at this point, as to whether the impacts of climate change can be said to have been clearly detected in the historical record for tropical cyclones, although some claims have been made with respect to the increasing intensity of the strongest storms [6,7].

In a recent paper, Kossin et al. [8] described an analysis based on the latest version of the ADT-HURSAT dataset [9,10] (where ADT stands for Advanced Dvorak Technique and HURSAT stands for HURricane SATellite), in which they showed that there is a statistically significant (outside a 2.5%–97.5% confidence interval—henceforth "significant") increase in the ratio of the number of category 3, 4 and 5 (cat35) fixes to the number of category 1, 2, 3, 4 and 5 (cat15) fixes in that dataset, over the period 1979 to 2017. Kossin et al. [8] used ADT-HURSAT, which is an objectively processed satellite dataset of global tropical cyclone observations, as it provides a more homogeneous time-series of tropical cyclone observations than the International Best Track Archive [11], even though the latter dataset may be more accurate in some ways. The fixes that they considered are 6 hourly measurements, and there are typically multiple fixes for each tropical storm. In general, cat35 fixes are defined as those that record windspeeds of 96+ knots, and cat15 fixes are those that record windspeeds of 64+ knots. Since in the ADT-HURSAT data intensity is discretized into 5 knot bins, this corresponds to windspeeds of 100+ knots and 65+ knots, respectively, in this dataset.

Kossin et al. [8] used a number of different statistical methods for investigating possible changes in the ratio of cat35 (henceforth major hurricane (MH)) to cat15 fixes (henceforth just "the MH ratio"), including consideration of the differences in the ratio between block averages of early and later periods of the data, and a linear trend analysis of the MH ratio combined into three-year sub-periods. These two methods both show significant signals corresponding to an increase in the MH ratio. They noted that this increase is consistent with theoretical and dynamical modelling results, which suggests an increase in the proportion of major tropical cyclones due to climate change [3].

However, from the point of view of tropical cyclone risk and risk modelling, the implications of this increase are hard to understand. The ratio of cat35 to cat15 fixes is not something that is explicitly included in most tropical cyclone risk models. In this article, our goal is therefore to analyze the increase identified in Kossin et al. [8] and understand what changes within the dataset are driving it, in terms of quantities that are more closely aligned with risk modelling. In particular, we will attempt to answer the following questions:

(a) is the increase in the MH ratio driven by changes in the numerator (the number of cat35 fixes), the denominator (the number of cat15 fixes), or both?

(b) is the increase in the MH ratio driven by any particular strength categories, or is it driven by several strength categories?

(c) is the increase driven by any particular regions, or is it the result of similar changes in all regions?

These are arithmetical questions about the increase in the MH ratio, rather than being questions about the physical drivers or mechanisms behind the increase. Answers to these questions can help risk modelers decide how they could adjust risk models to reflect the increasing MH ratio, should they wish to.

In Section 2 below, we provide a brief review of the linear trend analysis that Kossin et al. [8] used to identify the increase. In Section 3, we provide a heuristic analysis to understand what might be driving the linear trend. In Section 4, we provide a quantitative analysis using decomposition of the trend slope to identify the statistical drivers behind the linear trend. In Section 5, we summarize and draw some conclusions.

2. Dataset and Increase in the MH Ratio

In this section, we briefly review the increase in the ADT-HURSAT MH ratio that was identified in Kossin et al. [8]. The ADT-HURSAT data we use for our analysis are available on the internet as part of the supplementary information provided by Kossin et al. [8]. Figure 1a shows the annual numbers of category 1 and 2 (cat12), cat35, and cat15 fixes in the ADT-HURSAT dataset, for the period 1979 to 2017. In this and all the subsequent analysis, we omitted 1980 for reasons of incomplete data, following Kossin et al. [8]. In all the time series plots, linear interpolation was used between 1979 and 1981, although the calculation of trend-lines and significance levels all treat 1980 as a missing value. We see from Figure 1a that the three time series are correlated. That cat15 is correlated with both cat12 and cat35 is not surprising since the cat15 numbers are simply the sum of the cat12 and cat35 numbers. That cat12 and cat35 are correlated is also not surprising, since each storm that reaches cat35 must pass through cat12. We also see that the vertical distance between the cat12 and cat35 time series tends to decrease over time and is lower in the second half of the series than the first.

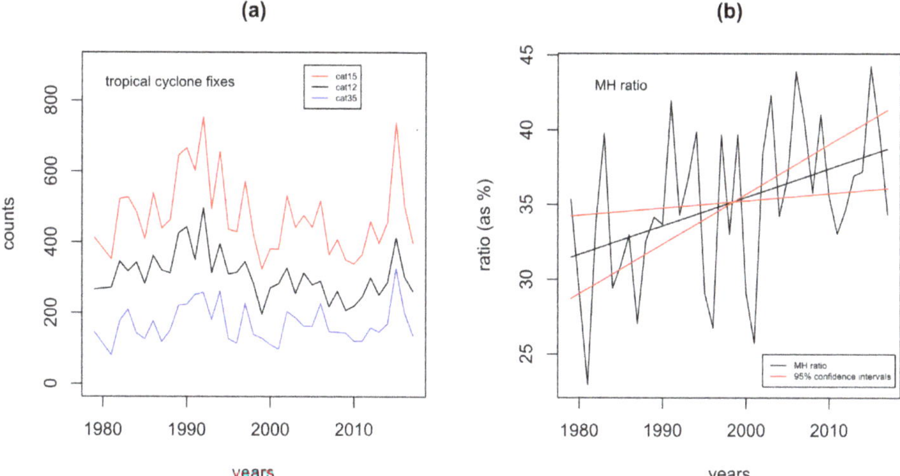

Figure 1. (**a**) shows numbers of cat15 (red), cat12 (black) and cat35 (blue) fixes from the ADT-HURSAT dataset, and (**b**) shows the ratio of cat35 to cat15 fixes (major hurricane (MH) ratio, black time series), with a fitted linear trend (black line), with lines showing 95% confidence limits (red). The linear trend is fitted using ordinary least squares (OLS) regression. The trend change is 7 percentage points over 39 years, and is significant with a *p*-value of 0.01. In the calculations, 1980 is treated as a missing value, and in the graph it is filled in using linear interpolation between 1979 and 1981.

Figure 1b shows the MH ratio. When the MH ratio is created, much of the variability cancels out between the two datasets because of the correlation between the series. Kossin et al. [8] also argued that this cancellation includes some cancellation of observational errors. An example would be that if a cat35 storm is missed (e.g., due to satellite failure), it will be missed from both the cat35 and cat12 records and hence will have a minimal impact on the MH ratio. This cancellation of errors is one of the reasons why Kossin et al. [8] chose to focus on the ratio of cat35 to cat15 fixes, rather than the numbers of fixes themselves.

However, from the point of view of trying to inform tropical storm risk quantification, it is necessary to consider the data for the actual counts of fixes, not just the MH ratio, and that is what we will do below. Using counts means that our analysis is exposed to a greater extent to the impacts of observational errors and variability in the data, because we lose the advantage of the cancellation of errors mentioned above. This should be taken into account in the interpretation of the results below: the count data should not be used for quantifying subtle signals. However, the main drivers we identify in our analysis below will turn out to be large signals that are hopefully reasonably robust.

If we write c_{12} as the number of cat12 fixes, c_{35} as the number of cat35 fixes, and c_{15} as the number of cat15 fixes then $c_{15} = c_{12} + c_{35}$ and the MH ratio r is given by

$$r = \frac{c_{35}}{c_{15}} = \frac{c_{35}}{c_{12} + c_{35}} = \frac{1}{1 + \frac{c_{12}}{c_{35}}} \tag{1}$$

That the cat12 and cat35 time series get closer together over time in Figure 1a implies that c_{12}/c_{35} is reducing, which in turn implies that r is increasing, which is what we see in Figure 1b.

One of the methods used by Kossin et al. [8] was to group the annual values of the MH ratio into triads (groups of three years), and consider statistical significance of a trend in the triad values, where the trend was fitted using Theil–Sen regression. As an alternative, and somewhat more simply, we fit a linear trend directly to the annual values shown in Figure 1b using ordinary least squares (OLS), with 1980 omitted. Using OLS facilitates the decomposition analysis we describe in Section 4 below. The OLS trend is also shown in Figure 1b. Above and below the OLS trend-line, we show two other lines which have the same mean as the OLS trend-line, but slopes which give the 95% confidence intervals around the trend slope. The OLS trend is positive and significant, with a p-value of $p = 0.01$.

Fitting a linear trend to these ratios is not strictly defensible, since the ratios are bounded between zero and one, while the trend line ultimately stretches from minus infinity to infinity. We tested the importance of avoiding this problem by transforming the ratios using a probit transform, to give them a possible range from minus infinity to plus infinity, before fitting the trend. This, however, made little difference to the results, since the range of the ratios is rather small, and the probit transform is close to linear over this range. For the remainder of our analysis, we proceed with the untransformed values for simplicity.

Based on Kossin et al.'s [8] analysis, and the linear trend analysis given above, there seems little doubt that the increase in the MH ratio is robust (i.e., is not an artefact of interannual variability in the dataset). However, from the point of view of fully understanding the implications of the increase for risk analysis, the question remains as to what is driving the increase.

3. Results

In this section, we perform a heuristic investigation to try to understand what might be driving the trend in the MH ratio. We will then complement this with a quantitative analysis in Section 4.

3.1. Numerator vs. Denominator

A first question one might ask about the cause of the trend shown in Figure 1b is: is it changes in the numerator of the ratio driving the trend, or the denominator of the ratio, or both? Figure 2a–c show the time series for the numerator (cat35 fixes), for cat12 fixes, and for the denominator (cat15 fixes), respectively, with linear trends fitted to all. The cat35 fixes show a very slight and not significant upward trend (p-value 0.79). The cat12 fixes show a significant downward trend (p-value 0.02). The cat15 fixes show a slight and not significant downward trend (p-value 0.23). Figure 2d shows the ratio of cat35 fixes to cat12 fixes, with a fitted trend, which has a positive slope. This is significant, with a p-value of 0.01. Table 1 summarizes the values of the significance levels of the main global trends in numbers of fixes for different strength categories.

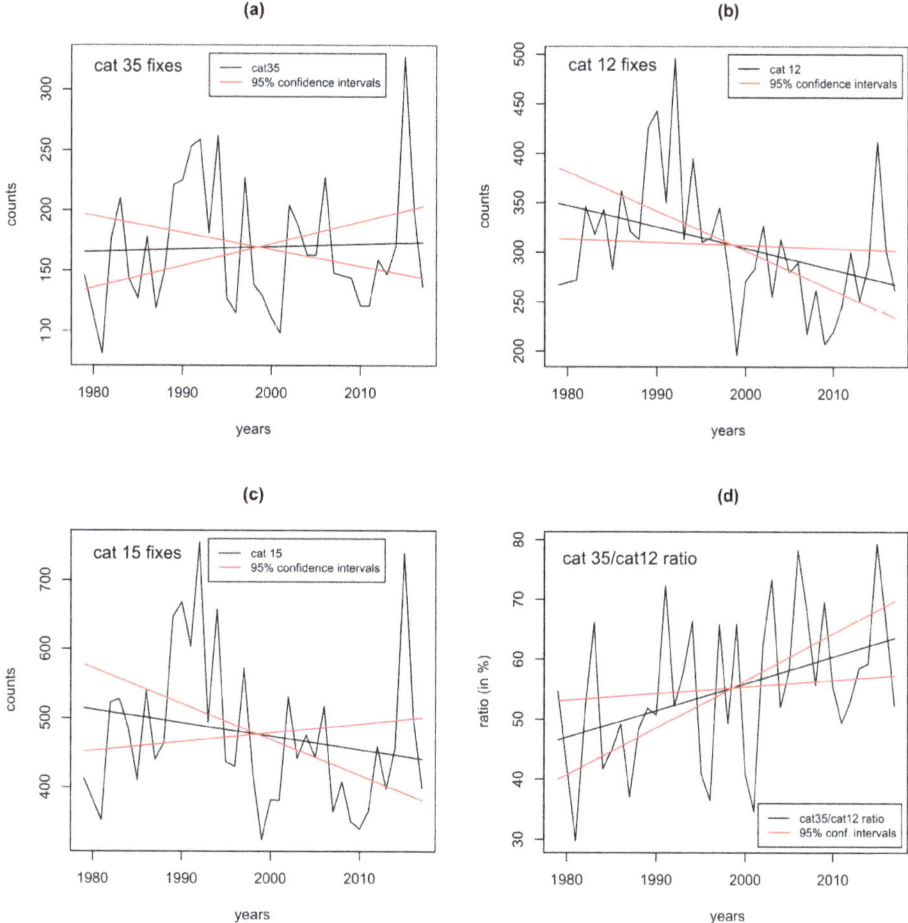

Figure 2. (**a**) shows numbers of cat35 fixes (black time series) (as in Figure 1a), with a linear trend (black line), and 95% confidence intervals trend (red lines). (**b**) shows cat12 in the same format; (**c**) shows cat 15 in the same format, and (**d**) shows the ratio of cat35 to cat12 in the same format.

Table 1. Trend slopes and significance levels for global trends in numbers of tropical cyclone fixes. The first column (Trend) shows trend slope values expressed as numbers of fixes per year. The second column (Significance) shows the p-value for significance of the trend.

	Trend	Significance
cat1	−1.95	0.004
cat2	−0.20	0.54
cat3	0.28	0.45
cat4	0.08	0.86
cat5	−0.14	0.18
cat12	−2.15	0.02
cat15	−1.94	0.23
cat35	0.21	0.79

Subject to quantitative confirmation in Section 4 below, we conclude from this that the upward trend in the MH ratio is to a small extent due to a non-significant upward trend in the cat35 fixes, but is mostly due to a downward trend in the cat15 fixes, which in turn is due to a significant downward trend in the cat12 fixes. Aspects of this downtrend were discussed in Holland and Bruyere [12].

3.2. Category of Storm Fixes

Given the above results, one might ask whether the upward trend in the cat35 fixes is due to cat3s, or cat4s, or cat5s, and whether the downward trend in cat12 fixes is due to cat1s or cat2s. Figure 3 shows trends for all five categories individually in sections (a) to (e). Trend slopes and p-values for the trends are given in Table 1. There is a strong and highly significant downward trend in the cat1 fixes (p-value 0.004), and a weak and not significant downward trend in the cat2 fixes. These two combine to give the downward trend in the cat12 fixes.

There are weak upward trends in the cat3 and cat4 fixes, and a downward trend in the cat5 fixes. None are significant but they combine to give the weak upward trend in the cat35 fixes. In the cat15 time series, the downward trend in cat12s is large enough to dominate the upward trend in the cat35s, to give a net downward trend overall. Combining this with the previous analysis leads us to conclude that the upward trend in the MH ratio is likely mostly driven by the significant downward trend in the cat1s.

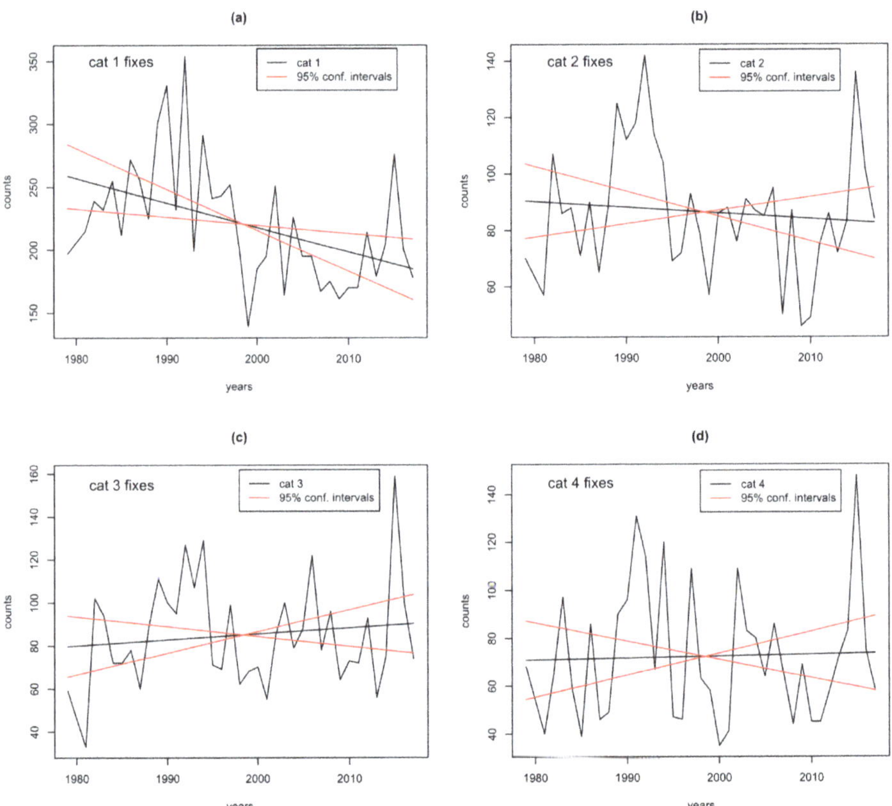

Figure 3. *Cont.*

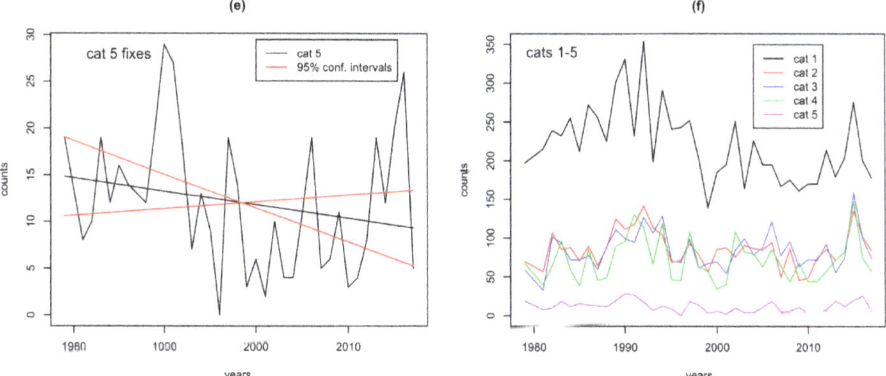

Figure 3. (**a**) shows the number of cat1 fixes (black time series), with a linear trend (black line), and lines showing 95% confidences on the trend slope; (**b–e**) show cat2, cat3, cat4 and cat5 fixes in the same format; (**f**) shows time series of all five categories: cat1 (black), cat2 (red), cat3 (blue), cat4 (green), cat5 (purple).

3.3. Regional Analysis

Based on the above analysis, one might also ask which regions are driving the changes in cat1 and cat34 fixes. The ADT-HURSAT data categorize each storm into one of seven regions: North Atlantic (NA), Eastern North Pacific (EP), Western North Pacific (WP), South Pacific (SP), Southern Indian Ocean (SI), Northern Indian Ocean (NI), and South Atlantic (SA). The South Atlantic region, however, only has one storm with cat15 fixes and so we will ignore it and only consider the other six regions.

Figure 4a shows the trends in the number of cat1 fixes for these six regions. Corresponding time series are shown in Figure A1. We see that there is a positive trend in the North Atlantic, and negative trends everywhere else. The only significant trends are the negative trends in the EP and SP basins. We conclude that the significant negative trend in global cat1 fixes, which is driving the negative trend in global cat12 fixes, and which appears to be the most important factor driving the positive trend in the MH ratio, comes from all basins except the North Atlantic.

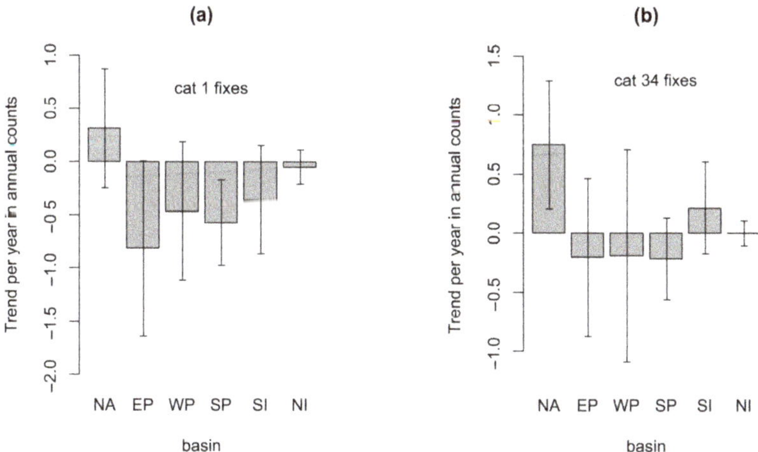

Figure 4. (**a**) shows the trends in the number of cat1 fixes by basin with 95% confidence intervals, and (**b**) shows the trends in the number of cat34 fixes by basin with 95% confidence intervals. The time series to which these trends relate are shown in Figures A1 and A2.

Figure 4b shows the trends in cat34 fixes by region. Corresponding time series are shown in Figure A2. We see that there is a significant positive trend in the North Atlantic, a non-significant smaller positive trend in the Southern Indian basin, essentially no trend in the Northern Indian basin, and non-significant negative trends in all other regions. We conclude that the positive trend in the global cat34 fixes, which is driving the positive trend in the global cat35 fixes, which is partly driving the positive trend in the MH ratio, comes almost entirely from the North Atlantic.

Overall, we can conclude, subject again to quantitative confirmation in Section 4 below, that the upward trend in the MH ratio seems to be driven mostly by a negative trend in cat1 fixes outside the North Atlantic, and to some extent by a positive trend in cat34 fixes in the Atlantic.

4. Quantitative Analysis

We now describe an analysis in which we decompose the trend in the MH ratio into various component terms, in order to make the heuristic analysis of the previous section quantitative.

4.1. Decomposition by Numerator and Denominator

In our first decomposition, we will decompose the trend in the MH ratio into a linear combination of trends in the numerator (cat35 fixes) and denominator (cat15 fixes), plus a residual. We will write β_r as the trend slope of the MH ratio, β_{35} as the trend slope of the number of cat35 fixes, β_{15} as the trend slope of the number of cat15 fixes and β_{12} as the trend slope of the number of cat12 fixes. In addition, we will write μ_{35} as the mean of the number of cat35 fixes, μ_{15} as the mean of the number of cat15 fixes, and $\mu_r = \mu_{35}/\mu_{15}$. Then, using the standard OLS formulae for all the trend slopes, it is relatively simple to show that:

$$\beta_r = \mu_r\left(\frac{\beta_{35}}{\mu_{35}} - \frac{\beta_{15}}{\mu_{15}}\right) + e \tag{2}$$

where e is a residual term involving an infinite expansion of non-linear terms. This equation matches intuition in the sense that if the trend in the cat35 fixes β_{35} increases then the trend in the MH ratio increases, while if the trend in the cat15 fixes β_{15} increases then the trend in the MH ratio decreases. One way to understand this equation is that it approximates the trend in the MH ratio, expressed as a proportion (β_r/μ_r), in terms of the trend in cat35 fixes, expressed as a proportion (β_{35}/β_{15}), minus the trend in the cat15 fixes, also expressed as a proportion (β_{15}/μ_{15}). Following the discussion about the impacts of observational errors in Section 2, the estimates of trends in the numbers of fixes in subsets of the data are likely more prone to the impacts of observational errors than the estimate of the trend in the MH ratio. In Equation (2), this may mean that there are errors in slopes β_{35} and β_{15} which cancel, to some extent, and hence contribute to less error in the slope β_r. As a result, and also because of general variability in the number of tropical storm fixes, the percentages in this and all the subsequent decompositions should not be considered as highly precise estimates.

We calculated all four terms in the above expression (Equation (2)) and evaluated them as a percentage of the trend in the MH ratio β_r. We find that the residual accounts for 1.0% of β_r, indicating that the trend in the MH ratio β_r is well approximated by a linear combination of the trends in the cat35 and cat15 fixes β_{35} and β_{15}. β_{35} accounts for 23.0% of β_r and β_{15} accounts for 76.0% of β_r.

This result confirms what we found in the heuristic analysis in the previous section: the upward trend in the MH ratio is driven mainly by the downward trend in the number of cat15 fixes, and partly by the upward trend in the number of cat35 fixes.

If instead we decompose the trend in the slope of the MH ratio β_r into the trend in the slope of the cat35 fixes β_{35} and the trend in the slope of the cat12 fixes β_{12}, we find that

$$\beta_r = \mu_r\left(\beta_{35}\left[\frac{1}{\mu_{35}} - \frac{1}{\mu_{15}}\right] - \frac{\beta_{12}}{\mu_{15}}\right) + e \tag{3}$$

The residual remains the same, while β_{35} now accounts for 14.9% of β_r and β_{12} accounts for 84.2% of β_r. This result again confirms what we saw in the previous section: the upward trend in the MH ratio is driven mainly by the downward trend in the number of cat12 fixes, and only partly by the upward trend in the number of cat35 fixes. The downward trend in the cat12 fixes is between five and six times more important than the upward trend in cat35 fixes in determining the upward trend in the MH ratio. We should also bear in mind that the upward trend in the cat35 fixes is not significant and may not be robust.

4.2. Decomposition by Category of Storm

We can decompose the trend in the MH ratio further into contributions from the trends in every category of storm strength. We write $\beta_1, \beta_2, \beta_3, \beta_4$ and β_5 as the OLS trends in the numbers of cat 1, 2, 3, 4 and 5 fixes. We can then decompose the MH ratio into components for each of these trend slopes and a non-linear residual as follows:

$$\beta_r = \mu_r\left([\beta_3 + \beta_4 + \beta_5]f - \frac{\beta_1 + \beta_2}{\mu_{15}}\right) + e \tag{4}$$

where $f = \left[\frac{1}{\mu_{35}} - \frac{1}{\mu_{15}}\right]$. Applying this new decomposition, we find percent contributions from the 5 categories from 1 to 5 of 76%, 8%, 20%, 5%, and −10% (remembering that this finer level of decomposition is even more prone to errors due to variability and observational error). We see that by far the biggest contribution to the upward trend in the MH ratio comes from the downward trend in cat1 fixes (76%). The second biggest contribution by strength category comes from the upward trend in cat3 fixes (20%). This confirms what we were able to deduce heuristically from Figure 3 in the previous section.

4.3. Decomposition by Region

We can extend the above decomposition analysis even further and decompose the trend in the MH ratio into contributions for each of the five strength categories and six basins, giving 30 categories in all. The results are shown in Table 2 as percentage contributions to the trend in the MH ratio. Sizes of trends for each of the 30 categories are shown in Table 3. The numeric values in the first two columns of Tables 2 and 3 are similar, but this is purely numeric coincidence. The percentage values in Table 2 add up to 102% because of rounding error. Combining the information from Tables 2 and 3, we see that the biggest contributors to the upward trend in the MH ratio are downward trends in cat1 fixes in the three Pacific basins and the Southern Indian Ocean, and upward trends in cat34 fixes in the North Atlantic. Cat35 fixes in the Pacific basins show downward trends, which contribute negatively to the trend in the MH ratio (i.e., tend to counteract the positive upward trend in the ratio), and cat1 fixes in the North Atlantic show a positive trend that also contributes negatively to the trend in the ratio.

Table 2. Percentage contributions to the trend in the MH ratio, by region and storm category.

	cat1	cat2	cat3	cat4	cat5	Total
NA	−12	−12	31	23	3	33
EP	32	6	−7	−7	−1	23
WP	18	5	2	−16	−12	−3
SP	23	8	−10	−5	−1	15
SI	14	3	5	10	1	33
NI	2	−1	0	0	0	1
total	77	9	21	5	−10	102

Table 3. Trends in the number of fixes, by category and region, in units of counts per 39 years.

	cat1	cat2	cat3	cat4	cat5	Total
NA	12	12	17	13	1	55
EP	−32	−6	−4	−4	−1	−47
WP	−18	−4	1	−9	−6	−36
SP	−22	−8	−6	−3	0	−39
SI	−14	−3	3	6	1	−7
NI	−2	1	0	0	0	−1
total	−76	−8	11	3	−5	−75

These results confirm what we previously deduced from Figure 4 in the previous section, which is that the upward trend in the MH ratio seems to be driven mostly by a negative trend in the cat1 fixes outside the North Atlantic (which is highly significant, with a p-value of 0.0035), and to some extent by a positive trend in cat34 fixes in the Atlantic (which is also significant, with a p-value of 0.01).

4.4. Regional Analysis of Ratios

Motivated by the results in Tables 2 and 3, Table 4 shows trends for various different versions of the cat35/cat15 ratio, and their significance. The first column shows the ratio for all basins together, as has been discussed in Section 2 above. This trend is significant, as we have seen above. The second column shows the trend in the cat35/cat15 ratio for all regions excluding the North Atlantic. The North Atlantic shows upward trends in both cat12s and cat35s, but the upward trend in the cat35s is larger (driven by a strong upward trend in cat34s), and hence removing the North Atlantic reduces the trend in the ratio, and the trend, although still positive, is no longer significant. We see that the strong upward trend in the North Atlantic cat34s is an essential component in the significance of the trend in the global MH ratio. The third column shows the North Atlantic on its own, which shows a large and significant upward trend in the ratio. Of the other five basins, four show positive trends in the ratio, of which one is significant (the Southern Indian ocean), although these regional trends, even more than the global trends, must be understood in the context of the large uncertainties in the data [4]. From Table 3, we see that the upward trends in the ratio in the three Pacific basins are driven by large decreases in the number of cat1s which are not offset by simultaneous increases in cat35s. The changes in these basins would seem to be best described as reductions in overall tropical cyclone frequency, with small changes in the distribution of intensities. With regard to the robustness of the trend in the global MH ratio, we can say that the trend is robust to a certain extent: after removing the North Atlantic, the sign of the trend remains the same, even though the significance is lost, and those of the other basins mostly show the same sign of trend. Given the level of variability of tropical cyclone numbers, and the size of the trend that would be expected from modelling studies [3], one would not expect the trend to be completely robust to all these tests.

Table 4. Fitted trends for various cat35/cat15 fixes ratios. The first row (Trend) shows trend values expressed as change in the ratio over 39 years. The second row (Significance) shows the p-value for significance of the trend. The first column (ALL) is for all regions combined. The second column (-NA) is for all regions excluding the North Atlantic.

	ALL	-NA	NA	EP	WP	SP	SI	NI
Trend	0.07	0.05	0.31	0.04	0.00	0.12	0.23	−0.04
Significance	0.0099	0.0910	0.0036	0.4782	0.9735	0.1576	0.0003	0.7083

4.5. Discussion

Combining the results from Section 3 above and from this section, we draw the following conclusions:

(1) The upward trend in the global MH ratio is mainly driven by a downward trend in the number of cat1 fixes in basins other than the North Atlantic.

(2) A secondary and less important driver is the upward trend in the number of cat34 fixes in the North Atlantic. Although this driver is less important than the decrease in the cat1 fixes, without it the trend in the MH ratio is no longer significant.

(3) Of the other five basins, only one shows a significant upward trend in the basin-level ratio (while noting that the trends in individual basins are subject to a greater extent to observational errors)

(4) The small upward trend in global total cat35 fixes shown in Figure 2a does not reflect a uniform global increase in cat35 fixes, but is driven by the large increase in the North Atlantic, an increase in the Southern Indian Ocean, and decreases in the three Pacific basins.

On a global aggregate level, we see a pattern consisting of a decrease in weak tropical cyclone fix numbers and an increase in strong tropical cyclone fix numbers. This is consistent with the idea of increasing intensity of individual tropical cyclones with climate change. However, this effect is mostly driven by combining together different behaviors in the North Atlantic and the other basins, and it is generally accepted that the large recent increase in North Atlantic hurricane numbers is likely strongly affected by regional climate variability as well as climate change. Possible explanations for the trends in the North Atlantic include variations in anthropogenic aerosols [13,14], natural climate variability on multidecadal time-scales [15], and the effects of volcanic eruptions [14,16]. As a result, it is difficult to determine accurately to what extent the increase in the MH ratio is due to climate change and to what extent the increase is due to regional climate variability.

5. Conclusions

In a study based on the ADT-HURSAT database of global tropical cyclone observations, Kossin et al. [8] identified a statistically significant increasing trend in the ratio of the number of cat35 fixes to the number of cat15 fixes (the MH ratio). We have explored the origins of this trend in terms of trends in the underlying numbers of fixes by storm category and by region in order to inform decisions about how tropical cyclone risk models might be adjusted to reflect the trend, for risk modelers interested in making such an adjustment.

We find that the main driver of the trend in the MH ratio is a downward trend in the cat15 fixes, which itself is driven by a statistically significant downward trend in the number of cat12 fixes, which itself is mostly driven by a statistically significant downward trend in the number of cat1 fixes. That in turn is driven by a statistically significant downwards trend in the number of cat1 fixes outside the North Atlantic basin. The North Atlantic cat1 fixes show an increasing trend, unlike any other basin.

The downward trend in cat1 fixes has an implication for regions affected by tropical cyclones, which is that the number of cat1 fixes occurring today in all regions except the North Atlantic seems to be lower than in previous decades. Risk estimates outside the North Atlantic based on the average number of cat1 fixes over previous decades might be too high.

A secondary driver for the upward trend in the MH ratio is a weak and non-significant upward trend in the cat 35 fixes. This is driven by a strong and significant upward trend in the number of cat34 fixes in the North Atlantic, which counteracts a downward trend in the number of cat35 fixes on average over the other basins. The trend in the number of cat34 fixes also has implications for risk estimation. Risk estimates for the North Atlantic based on averages of numbers of cat34 fixes over previous decades might be too low.

All these results must be understood in the context of the uncertainties in the underlying data. For instance, as discussed in Kossin et al. [8], there are gaps in the data. While these gaps do not have a big impact on estimates of the MH ratio and the results presented in Kossin et al., they are likely to have a bigger impact on the estimates of trends in the numbers of fixes that we have presented here.

Based on the above results, there are various ways that risk modelers could make adjustments if they wished to incorporate the trend in the MH ratio into their models. One way would be to adjust the global numbers of cat35 and cat12 fixes uniformly across all basins, although this would ignore the strong regional signals described above. Another way would be to adjust the numbers of cat34 fixes in the Atlantic basin, and the numbers of cat1 fixes in other basins, since these are the regional signals that are the main drivers for the increase. Given the uncertainties, all such adjustments to risk models should perhaps be treated as exploratory sensitivity tests, to test the impact of different hypotheses, rather than as fully credible views of hurricane risk.

Author Contributions: Conceptualization, S.J.; methodology, S.J.; software, S.J.; validation, S.J., N.L.; formal analysis, S.J.; writing—original draft preparation, S.J.; writing—review and editing, S.J., N.L.; visualization, S.J. All authors have read and agreed to the published version of the manuscript.

Funding: This research received no external funding. The work was funded by the authors.

Acknowledgments: We would like to acknowledge the help of Jim Kossin, and the very helpful suggestions from three anonymous reviewers which greatly improved the manuscript.

Conflicts of Interest: The authors declare no conflict of interest.

Appendix A

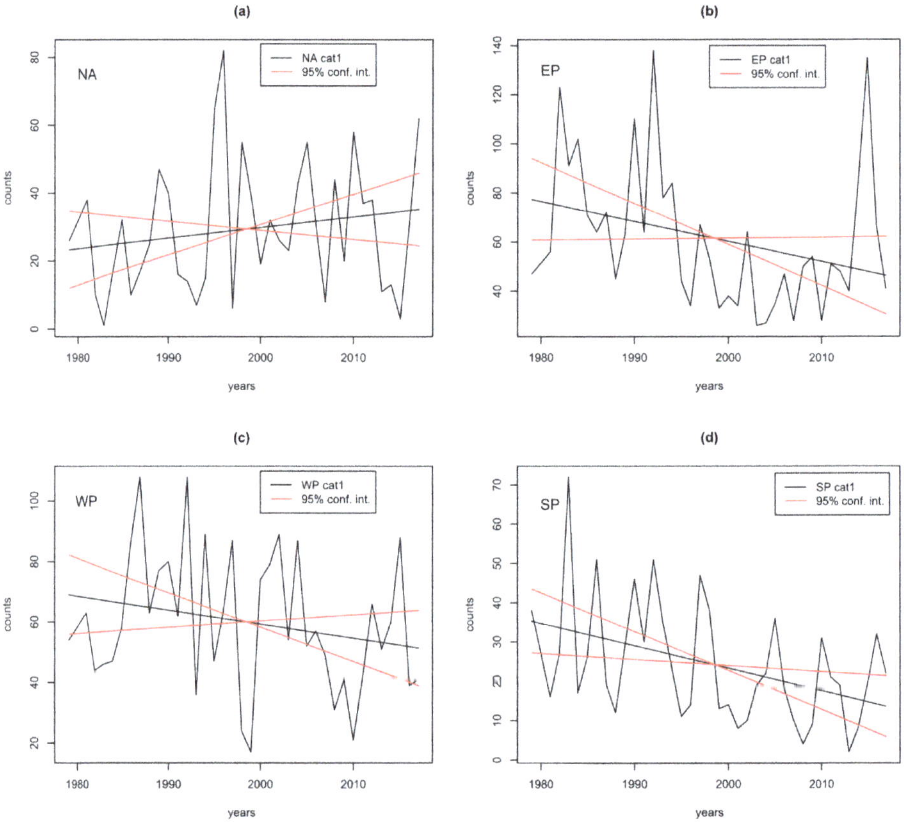

Figure A1. *Cont.*

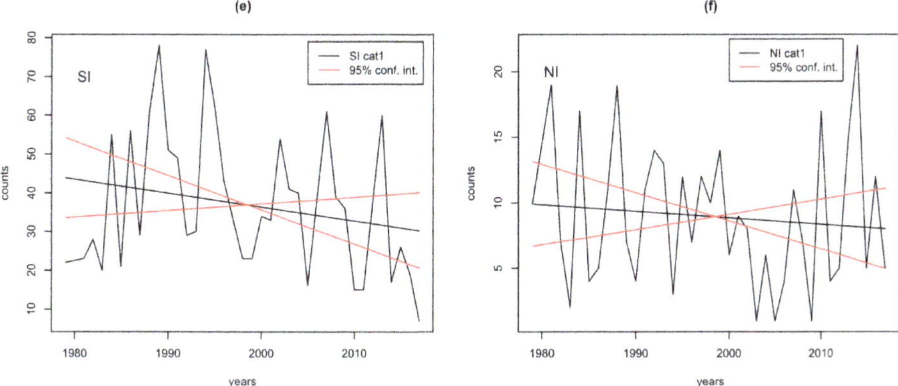

Figure A1. (**a**–**f**) show the number of cat1 fixes (black time series), with a linear trend (black line), and lines showing 95% confidence intervals on the trend slope, for 6 basins.

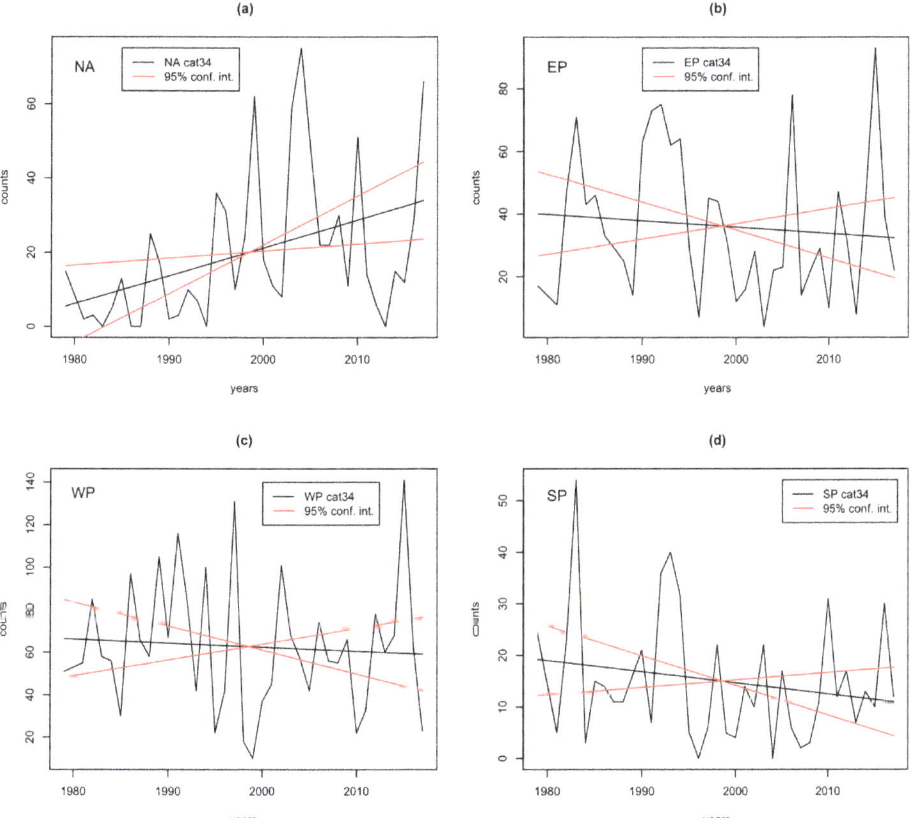

Figure A2. *Cont.*

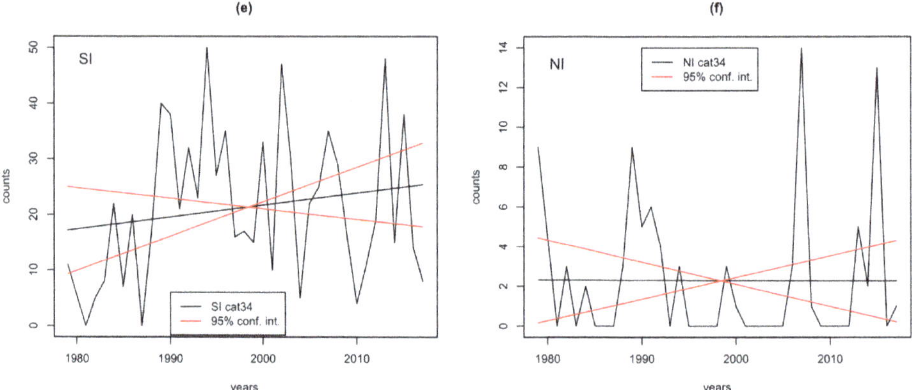

Figure A2. (**a–f**) show the number of cat34 fixes (black time series), with a linear trend (black line), and lines showing 95% confidence intervals on the trend slope for 6 basins.

References

1. Emanuel, K. The dependence of hurricane intensity of climate. *Nature* **1987**, *326*, 483–485. [CrossRef]
2. Sobel, A.; Camargo, S.; Hall, T.; Lee, C.; Tippett, M.; Wing, A. Human influence on tropical cyclone intensity. *Science* **2016**, *353*, 242–246. [CrossRef] [PubMed]
3. Knutson, T.; Camargo, S.; Chan, J.; Emanuel, K.; Ho, C.; Kossin, J.; Mohapatra, M.; Satoh, M.; Sugi, M.; Walsh, K.; et al. Tropical cyclones and climate change assessment: Part II: Projected response to anthropogenic warming. *Bull. Am. Meteorol. Soc.* **2020**, *101*, E303–E322. [CrossRef]
4. Landsea, C.; Harper, B.; Hoarau, K.; Knaff, J. Can we detect trends in extreme tropical cyclones? *Science* **2006**, *313*, 452–454. [CrossRef] [PubMed]
5. Goldenberg, S.; Landsea, C.; Mestas-Nunez, A.; Gray, B. The Recent Increase in Atlantic Hurricane Activity: Causes and Implications. *Science* **2001**, *293*, 474–479. [CrossRef] [PubMed]
6. Elsner, J.; Kossin, J.; Jagger, T. The increasing intensity of the strongest tropical cyclones. *Nature* **2008**, *455*, 92–95. [CrossRef] [PubMed]
7. Knutson, T.; Camargo, S.; Chan, J.; Emanuel, K.; Ho, C.; Kossin, J.; Mohapatra, M.; Satoh, M.; Sugi, M.; Walsh, K.; et al. Tropical cyclones and climate change assessment: Part 1: Detection and attribution. *Bull. Am. Meteorol. Soc.* **2019**, *100*, 1987–2007. [CrossRef]
8. Kossin, J.; Knapp, K.; Olander, T.; Velden, C. Global Increase in major tropical cyclone exceedance probability over the past four decades. *Proc. Natl. Acad. Sci. USA* **2020**, *117*, 11975–11980. [CrossRef] [PubMed]
9. Kossin, J.; Olander, T.; Knapp, K. Trend analysis with a new global record of tropical cyclone intensity. *J. Clim.* **2013**, *26*, 9960–9976. [CrossRef]
10. Knapp, K.R.; Kossin, J.P. New global tropical cyclone data from ISCCP B1 geostationary satellite observations. *J. Appl. Remote Sens.* **2007**, *1*, 013505.
11. Knapp, K.; Kruk, K.; Levinson, D.; Diamond, H.; Neumann, C. The international best track archive for climate stewardship (IBTrACS): Unifying tropical cyclone best track data. *Bull. Am. Meteorol. Soc.* **2010**, *91*, 363–376. [CrossRef]
12. Holland, G.; Bruyere, C. Recent intense hurricane response to global climate change. *Clim. Dyn.* **2014**, *42*, 617–627. [CrossRef]
13. Dunstone, N.J.; Smith, D.M.; Booth, B.B.; Hermanson, L.; Eade, R. Anthropogenic aerosol forcing of Atlantic tropical storms. *Nat. Geosci.* **2013**, *6*, 534–539. [CrossRef]
14. Murikami, H.; Delworth, T.L.; Cooke, W.F.; Zhao, M.; Xiang, B.; Hsu, P.C. Detected climatic change in global distribution of tropical cyclones. *Proc. Natl. Acad. Sci. USA* **2020**, *117*, 10706–10714. [CrossRef] [PubMed]

15. Yan, X.; Zhang, R.; Knutson, T. The role of Atlantic overturning circulation in the recent decline of Atlantic major hurricane frequency. *Nat. Commun.* **2017**, *8*, 1695. [CrossRef] [PubMed]
16. Pausata, F.S.; Camargo, S.J. Tropical cyclone activity affected by volcanically induced ITCZ shifts. *Proc. Natl. Acad. Sci. USA* **2019**, *116*, 7732–7737. [CrossRef] [PubMed]

Publisher's Note: MDPI stays neutral with regard to jurisdictional claims in published maps and institutional affiliations.

Article

Future Changes in Western North Pacific Tropical Cyclone Genesis Environment in High-Resolution Large-Ensemble Simulations

Hironori Fudeyasu [1,*], Kohei Yoshida [2] and Ryuji Yoshida [3,4,5]

[1]   College of Education, Yokohama National University, Yokohama 240-8501, Japan
[2]   Meteorological Research Institute, Tsukuba 305-0052, Japan; kyoshida@mri-jma.go.jp
[3]   CIRES, University of Colorado Boulder, NOAA Chemical Sciences Laboratory, Boulder, CO 80305, USA; ryuji.yoshida@noaa.gov
[4]   RIKEN Center for Computational Science, Kobe 650-0047, Japan
[5]   Research Center for Urban Safety and Security, Kobe University, Kobe 657-8501, Japan
*   Correspondence: fude@ynu.ac.jp; Tel.: +81-45-339-3346

Received: 14 October 2020; Accepted: 10 December 2020; Published: 18 December 2020

Abstract: This study applied the database for Policy Decision making for Future climate change (d4PDF) and tropical cyclone (TC) genesis (TCG) environment factors to project future changes in the frequency and characteristics of TCs over the western North Pacific. We examined current and future TCG environmental conditions in terms of the contribution of five factors: shear line (SL), confluence region (CR), monsoon gyre, easterly wave (EW), and Rossby wave energy dispersion from a preexisting TC (PTC). Among summer and autumn TCs, the contributions of SL and EW to future TCG increased by about 4% and 1%, respectively, whereas those of CR and PTC decreased by the same amounts. In future climate projections, the average lifetime maximum intensity (LMI) of TCs associated with EW (EW-TCs) was significantly higher than those of TCs associated with other factors except PTC. At higher sea surface temperatures and wetter conditions, higher lower-tropospheric relative vorticity was related to increases in the development rate of EW-TCs. Findings of this study suggest that increases in the average LMI of all future TCs were caused by large contributions from the average LMI of future EW-TCs.

Keywords: tropical cyclone genesis environment; tropical cyclone projection; western North Pacific tropical cyclone

---

1. Introduction

Many studies have reported projections of tropical cyclone (TC) activity, including the frequency of TC genesis (TCG) and lifetime maximum intensity (LMI), under future global warming using high-resolution global models. One such study, Yoshida et al. [1], applied a large-scale ensemble dataset based on a high-resolution global atmospheric model the database for Policy Decision making for Future climate change (d4PDF) to assess the uncertainty of atmospheric internal variations and sea surface temperature (SST) increases; the frequency of global TCG was projected to decrease significantly and the proportion of intense TCs to increase by the end of the 21st century, based on the Representative Concentration Pathway 8.5 (RCP8.5) scenario. Over the western North Pacific (WNP), the TCG frequency was projected to decrease by about 40%, and the average LMI of TCs to increase by about 10%.

Several other TC projection studies have shown variant results. A review of Knutson et al. [2] that have projected future TC activity found a medium to high confidence for projected LMI increases, but less consensus for future TCG frequency changes. For example, Roberts et al. [3] projected a

10–20% decrease in TCG frequency using a high-resolution atmospheric model Met. Office Unified Model Global Atmosphere 3 configuration for RCP8.5 by the end of the 21st century. Knutson et al. [4] simulated a 35% decrease in TCG frequency using the high-resolution atmospheric model (HiRAM) and the Geophysical Fluid Dynamics Laboratory (GFDL) hurricane model for Representative Concentration Pathway 4.5 (RCP4.5). In contrast, the atmosphere–ocean-coupled high-resolution forecast-oriented low ocean resolution (HiFLOR) model [5] showed an increase in TCG frequency of 5.8%. A recent study by Zhang et al. [6] used a TC downscaling model and found that the Pacific TCs are going to become more frequent and more intense. Thus, future TC activity remains highly uncertain, and it is difficult to design measures such as social infrastructure to adapt to anticipated increases in TC variability. To reduce the uncertainty of TC activity projections, it is necessary to develop a comprehensive understanding of the detailed mechanisms of TCs in the models.

To assess modeled TCG, it is important to understand the environmental factors that influence TCG over the WNP. Ritchie and Holland [7] examined five large-scale flow patterns as factors contributing to TCG: shear line (SL), confluence region (CR), monsoon gyre (GY), easterly wave (EW), and Rossby wave energy dispersion from a preexisting TC (PTC). The SL is enhanced by horizontal cyclonic shear over the WNP. The CR is a confluence zone situated between the easterly trade winds and westerly monsoon winds at the eastern extremity of the Asian monsoon. The GY is a synoptic-scale gyre embedded within a developed Asian monsoon trough [8,9]. Thus, these factors are closely related to Asian monsoons over the WNP. The EW is related to the synoptic-scale trade easterly wind system; the trough of easterly trade winds provides an environment favorable for TCG [10,11]. A mature TC disperses its energy as a Rossby wave in a southeastward direction; low-pressure areas of the wave train sometimes develop into TCs [12–14].

Yoshida and Ishikawa [15] developed a TCG score detection method (TGS) as an objective index to identify the main environmental factors contributing to TCG. Using the TGS and data collected during the Japanese 25-year Reanalysis Project [16], Fudeyasu and Yoshida [17] determined the statistical characteristics of TCs over the WNP, stratified by five TCG environmental factors in the summer and autumn during 1979–2013. TCs that formed in large-scale flows mainly associated with a GY (hereafter, GY-TCs, and similar terms indicate a TC that formed in a large-scale flow mainly associated with an GY or similar factor) tended to develop slowly. The average LMI of PTC-TCs is larger than those of other factors, due to their longer development stage. On average, TC size of GY-TCs is larger at the TCG time than other TCG factors, whereas those of EW-TCs and PTC-TCs are smaller. Fudeyasu and Yoshida [17] summarized differences in the TCG factors that lead to characterization of TCs.

Previous studies [18,19] have projected the environmental factors on TCG over the WNP under future climate change in terms of genesis potential index or similar indices. None has explored future changes in the five TCG environment factors identified by Ritchie and Holland [7]. Assuming that five TCG factors are also the main contributors to TCG over the WNP, future changes in TC characteristics may be projected in terms of TCG frequency or the average LMI of future TCs. To enhance our understanding of TC activity under future climate change, the present study used the d4PDF large-scale ensemble dataset and TGS to investigate the TCG environment statistically under future climate simulations and stratified changes in TC characteristics by the TCG factors.

2. Methodology

In this study, we used the d4PDF large-scale ensemble simulation dataset for current and future climates. The d4PDF dataset was produced by the Meteorological Research Institute–Atmospheric General Circulation Model Version 3.2 (MRI–AGCM3.2H), which is based on a numerical weather prediction model of the Japan Meteorological Agency (JMA), with a horizontal resolution of 60 km [20]. The model has 64 vertical layers, and the model top is set at 0.01 hPa. Current (future) climate simulations of the d4PDF dataset were achieved using 100 (90) ensemble members based on initial atmospheric perturbations and short-term monthly SST perturbations. Six SST warming patterns used in future climate simulations are projected by six coupled models: the Community Climate System Model, version 4 (CCSM4), the Geophysical Fluid Dynamics Laboratory Climate Model, version 3 (GFDL-CM3), the Hadley Centre Global Environment Model, version 2—Atmosphere and Ocean (HadGEM2-AO), the Model for Interdisciplinary Research on Climate, version 5 (MIROC5), the Max Planck Institute Earth System Model, medium resolution (MPI-ESM-MR), and the Meteorological Research Institute Coupled Atmosphere–Ocean General Circulation Model, version 3 (MRI-CGCM3). For the RCP8.5 scenario, SST warming patterns are calculated as monthly differences in SSTs between 1991–2010 and 2080–2099. Climatological differences are scaled by a multiplier, such that MRI–AGCM3.2H reproduces surface air warming of 4 K above pre-industrial levels (+4 K). Current climate simulations were integrated for 60 years (1951–2010) using observed SSTs and sea ice concentrations obtained from the Centennial in situ Observation-Based Estimates (COBE) SST2 product [21]. Future climate simulations were performed for a 60-year period with constant warming roughly corresponding to the levels projected for 2090 under the RCP8.5 scenario adopted in Phase 5 of the Coupled Model Intercomparison Project (CMIP5). For each of the six warming SST patterns, 15-member ensemble experiments were conducted with adding small SST perturbations, giving a total of 90 members (see Mizuta et al. [22] in detail). Thus, the ensemble datasets for present and +4 K future climate scenarios are summarized as follows: 1951–2010 × 100 members (a 6000-year dataset) for the current climate, and 2051–2110 × 90 members (a 5400-year dataset) for the future climate. Further details of the experimental design were provided in Mizuta et al. [22].

In this study, we analyzed 180,753 (87,746) TCs in current (future) climate simulations of the d4PDF dataset. TC track data for the d4PDF dataset were derived from Yoshida et al. [1], which applied the same TC tracking method and threshold values as Murakami et al. [23], which reported that MRI–AGCM3.2H reproduced realistic TC activity in all aspects except for intensity. The TC tracking method and threshold values are provided in Yoshida et al. [1] and Murakami et al. [23].

This study applied the TGS developed by Yoshida and Ishikawa [15] to estimate the contributions of each TCG factor [7] for wind speeds at 850 hPa and sea level pressure, derived from d4PDF. The location and time of the first TC track data entry were used as the TCG location and time. The maximum contribution score was determined as the major flow pattern that mainly affected the TCG, among the SL, CR, GY, EW, and PTC. When a contributing factor could not be detected by the TGS algorithm, the flow pattern was categorized as unclassified flow (UCF). Details of the objective TGS algorithm are provided in previous studies [15,17,24].

This study performed statistical analysis of the following environmental conditions related to TCG, as derived from the d4PDF dataset: SST, magnitude of vertical shear (SHR), middle-tropospheric humidity (HMD), and lower tropospheric relative vorticity. SHR was defined as the difference in horizontal wind speed between heights of 200 and 850 hPa. HMD was defined as the specific humidity at a height of 500 hPa. The 850-hPa relative vorticity (VOR850) was calculated to represent lower tropospheric relative vorticity. These physical parameters were averaged over a 5° × 5° box relative to the TC center.

3. Results

3.1. Characteristics of Current TCs Stratified by TCG Environmental Factors

The annual average TCG over the WNP in the current d4PDF climate simulations was 30.1, which was slightly larger than the observed average of 26.2, obtained from the best-track archives of the Regional Specialized Meteorological Centers Tokyo–Typhoon Center. Figure 1a shows the seasonal changes in TCs stratified by the five TCG environmental factors and UCF. SL-TCs occurred throughout the year. The number of CR-TCs and GY-TCs tended to increase in summer. PTC-TCs tended to occur in autumn, when many TCs formed at lower latitudes. The seasonality of TCs, stratified by TCG environmental factors, was similar to that reported in Yoshida and Ishikawa [15], which considered TCG factors for all TCs, including tropical disturbances.

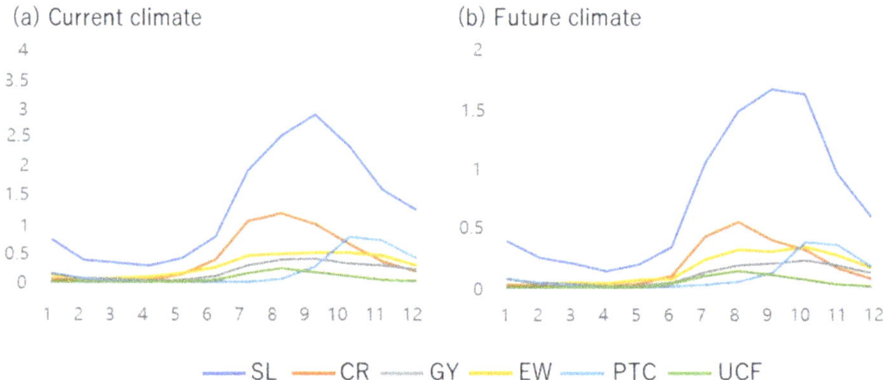

Figure 1. Monthly annual number of tropical cyclones (TCs), stratified by five flow patterns and unclassified flow (UCF) for (**a**) current and (**b**) future climate simulations performed using the d4PDF dataset. SL, shear line; CR, confluence region; GY, monsoon gyre; EW, easterly wave; PTC, Rossby wave energy dispersion from a preexisting TC.

To compare the characteristics of TCs accurately among TCG factors, we considered TCs in summer and autumn from June to November; all of the examined environmental factors occurred during this period. Among the 143,734 TCs that occurred in summer and autumn (Figure 2a), the most frequent were SL-TCs (72,673 TCs, 50.6% of the total), followed by CR-TCs (27,912 TCs, 19.4%), and EW-TCs (16,113 TCs, 11.2%). PTC-TCs (11,257 TCs) and GY-TCs (10,970 TCs) accounted for 7.8% and 7.6% of all TCs, respectively. The least frequent TCs were those for which the major contributing factors could not be determined, UCF-TCs (4809 TCs, 3.3%).

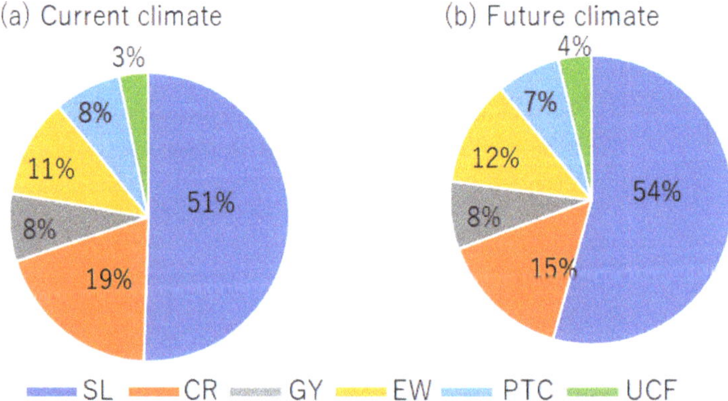

Figure 2. Distribution of frequency of TC genesis (TCG) environments stratified by five TCG factors and UCF from June to November over the analysis period for (**a**) current and (**b**) future climate simulations performed using the d4PDF.

Using TC data derived from the International Best Track Archive for Climate Stewardship (IBTrACS) [25] and Japanese 25-year Reanalysis Project [16], Fudeyasu and Yoshida [17] analyzed TCG frequency over the WNP, stratified by TCG factors. Among the 677 TCs that occurred in June to November during the analysis period (1979–2013), the most frequent were SL-TCs (300 TCs, 44.3% of the total), followed by CR-TCs (130 TCs, 19.2%), EW-TCs (92 TCs, 13.6%), GY-TCs (61 TCs, 9.0%),

PTC-TCs (66 TCs, 9.7%), and UCF-TCs (28 TCs, 4.1%). Although the proportion of SL-TCs (EW-TCs and PTC-TCs) in the d4PDF dataset was slightly higher (lower), the TCG frequency stratified by five TCG factors and UCF in the present study was almost the same as those reported in Fudeyasu and Yoshida [17].

This study performed statistical analysis of TCs stratified by TCG factors in current d4PDF climate simulations. For each factor that influences TCG, the relationship between the average contribution of each factor and those of all other factors was assessed using Student's *t*-test at a significance level of $p < 0.01$. For example, average characteristics of SL-TCs were compared to those of CR-TCs, GY-TCs, EW-TCs, PTC-TCs, and UCF-TCs.

Table 1 show the average locations of TCG for all TCs (Figure 3) and TCG factors (Figure 4) from June to November. There were significant differences in average TCG location among all factors. On average, SL-TCs and GY-TCs occurred significantly farther to the west than the other TCG factors ($p < 0.01$). On average, CR-TCs, EW-TCs, and UCF-TCs (GY-TCs, and PTC-TCs) occurred farther to the north (south) than other TCG factors.

Table 1. Statistical summary of the tropical cyclone (TC) characteristics associated with TCG environmental factors from June to November over a 6000-year analysis period in current climate simulations using the d4PDF dataset. Significant differences are indicated in bold and italic fonts (Student's *t*-test; $p < 0.01$). ALL, averaged for all TCs; SL, shear line; CR, confluence region; GY, monsoon gyre; EW, easterly wave; PTC, Rossby wave energy dispersion from a preexisting TC; UCF, unclassified flow.

	ALL	SL	CR	GY	EW	PTC	UCF
Longitude at TCG time (°N)	137.4	*135.2*	*141.0*	*132.9*	*140.6*	*141.4*	*138.8*
Latitude at TCG time (°E)	16.6	*16.6*	*17.2*	*16.3*	*17.7*	*13.5*	*17.6*
TC size at TCG time (km)	333.5	*343.2*	*321.7*	*349.2*	*317.0*	*262.1*	*358.6*
Development rate (m s^{-1} 6 h^{-1})	0.61	*0.63*	*0.56*	*0.55*	0.60	*0.69*	*0.55*
Movement speed (km h^{-1})	16.0	*15.4*	*16.6*	*14.7*	*17.5*	*17.6*	*15.6*
Duration of development (day)	3.1	3.2	*3.3*	*2.9*	*3.0*	3.3	*2.8*
LMI (m s^{-1})	25.9	*26.4*	*25.4*	*23.4*	25.8	*27.8*	*22.4*

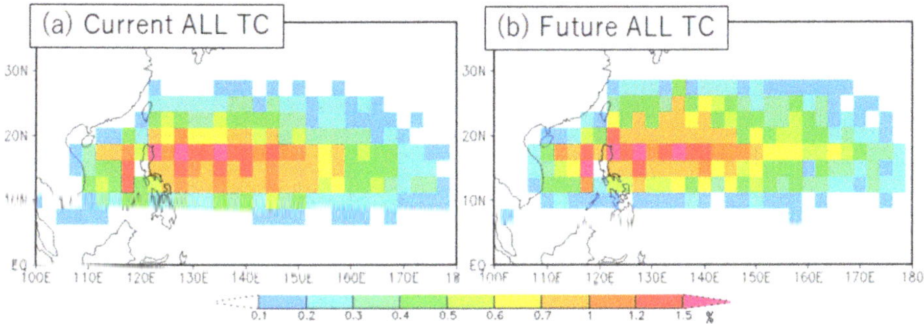

Figure 3. Distribution rate of TCG location, expressed as percentage values of the total number of TCG for (**a**) current all TCs and (**b**) future all TCs into a 2.5° × 2.5° latitude-longitude grid box with respect to the total number of TCG for (**a**) current all TCs and (**b**) future all TCs formed in the WMP from June to November over the analysis period in current and future climate simulations performed using the d4PDF.

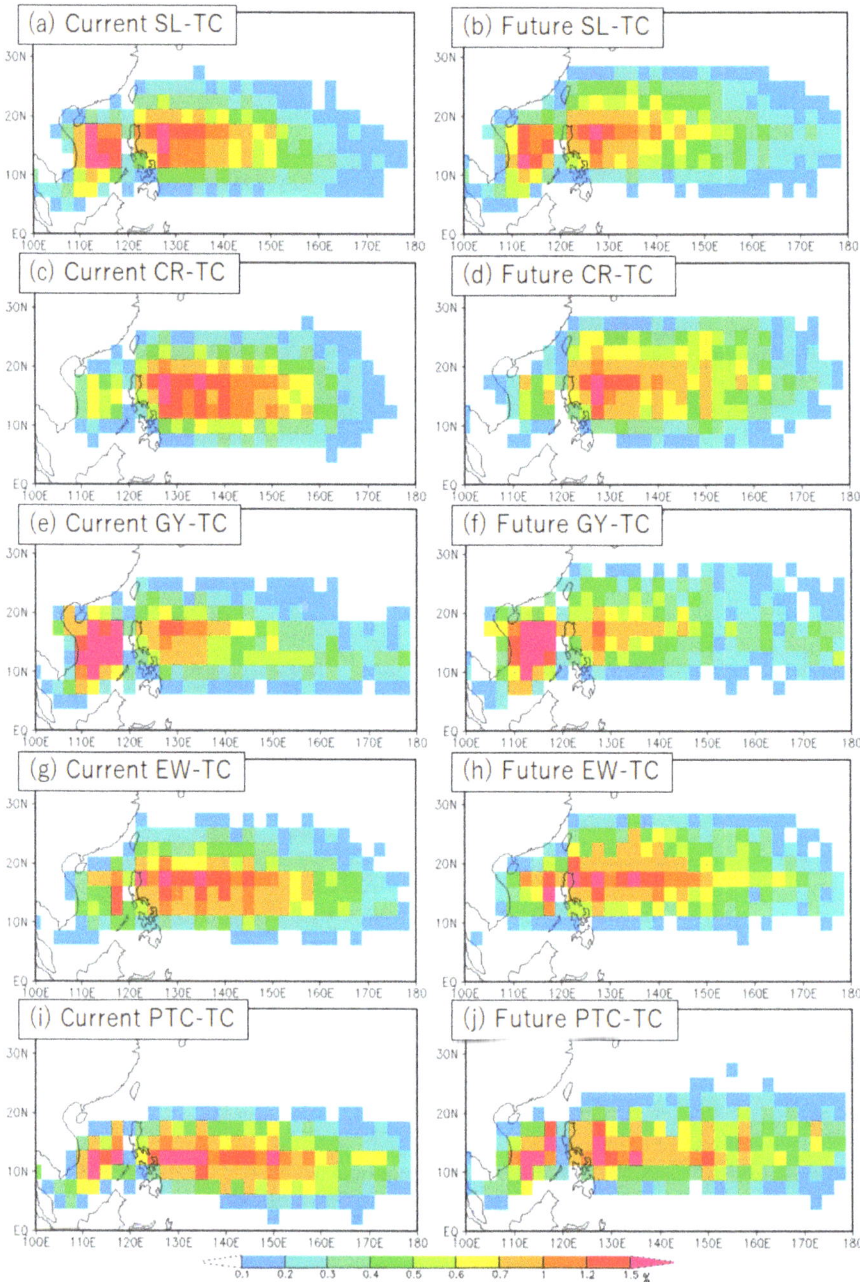

Figure 4. Same as Figure 3, except distribution rate (percent) of TCG location for (**a**) current SL-TCs, (**b**) future SL-TCs, (**c**) current CR-TCs, (**d**) future CR-TCs, (**e**) current GY-TCs, (**f**) future GY-TCs, (**g**) current EW-TCs, (**h**) future EW-TCs, (**i**) current PTC-TCs, and (**j**) future PTC-TCs.

Fudeyasu and Yoshida [17] clarified the characteristics of observed TCs in summer and autumn from 1979 to 2013, stratified by five TCG factors and UCF. The average TCG locations differed significantly among all factors ($p < 0.05$). The average location of EW-TCs (PTC-TCs) was farther east and north (south), whereas that of SL-TCs was farther west than those of all other TCG factors, which is consistent with our results (Table 1); we observed significant differences in average location for CR-TCs, GY-TCs, and UCF-TCs for current d4PDF climate simulations, due to very large sample sizes.

We defined TC size as the horizontal extent of TCs at the time of TCG; TC size was calculated as the maximum radius of the 1000-hPa contour from the TC center (Table 1). On average, GY-TCs were larger than TCs of all other factors, whereas EW-TCs and PTC-TCs were smaller (Table 1), which was consistent with the observations of Fudeyasu and Yoshida [17]. In the current d4PDF climate simulations, SL-TCs and UCF-TCs (CR-TCs) were larger (smaller) than TCs of the other factors.

The development rate at TCG, defined as the increase in maximum wind speed 6 h after the time of TCG, differed significantly among factors. GY-TCs had lower development rates than all other TCG factors, which was consistent with the observations reported in Fudeyasu and Yoshida [17]. In the current d4PDF climate simulations, the average development rates of SL-TCs and PTC-TCs (CR-TCs and UCF-TCs) were higher (lower) than those of the other TCG factors.

Table 1 also shows the average TC movement speed during the development stage, which is the duration from the time of TCG to maturity. The mature time is defined as the time at which maximum wind speed occurred. The movement speed was calculated in terms of the hourly horizontal displacement of the TC. The average movement speeds of CR-TCs, EW-TCs, and PTC-TCs were higher than those of the other factors in the development stage. The average duration of the development stage was longer for PTC-TCs than for the other TCG factors, which was consistent with Fudeyasu and Yoshida [17]. In the current d4PDF climate simulations, the average development stage of CR-TCs was longer than those of the other factors.

Highly significant differences in LMI among TCG factors were observed (Table 1 and Figure 5). PTC-TCs (UCF-TCs) tended to have higher (lower) LMI, which was consistent with Fudeyasu and Yoshida [17]. In the current d4PDF climate simulations, the average LMI of SL-TCs (CR-TCs and GY-TCs) was significantly higher (lower) than those of the other TCG factors ($p < 0.01$).

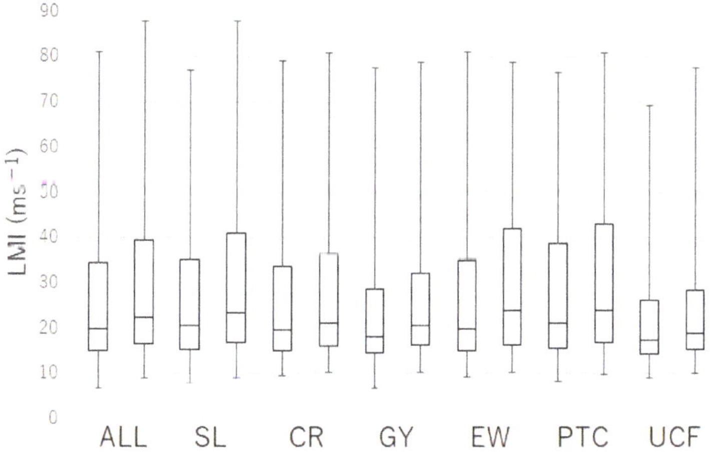

Figure 5. Boxplots of TC lifetime maximum intensity (LMI) stratified by TCG factors and UCF from June to November over the analysis period in current and future climate simulations performed using the d4PDF dataset. For each factor, boxes on the left and right are derived from current and future climate simulations, respectively. Vertical bar indicates the full range of the distributions; box indicates the interquartile range of 25–75% of all cases; bold horizontal line indicates the median.

3.2. Characteristics of Future TCs Stratified by TCG Environmental Factors

The annual average of TCG over the WNP in the future climate simulations was 16.2, while that in the current climate was 30.1. It was almost a half of the number in the current climate simulations, and this significant decrease was consistent with a future trend of TCG frequency reported in previous studies [26,27]. Figure 1b shows the seasonal changes in TCs stratified the five TCG factors and UCF; the seasonal changes due to each TCG factor were very similar to those shown in the current climate simulations. Among the 70,920 TCs projected to occur in summer and autumn in future climate simulations, the most frequent were SL-TCs (38,584 TCs, 54.4% of the total; Figure 2b). The percentage of SL-TCs was about 3.8% higher than that of the current climate simulations. The percentage of CR-TCs (10,666 TCs) was 15.0%, about 4.4% lower than that of the current climate simulations. The percentages of EW-TCs (8420 TCs, 11.9%) and PTC-TCs (5200 TCs, 7.3%) were about 1% higher and lower than those of the current climate, respectively. The percentages of GY-TCs (5404 TCs, 7.6%) and UCF-TCs (2646 TCs, 3.7%) were almost same as those of the current climate simulations.

The averages characteristics of all TCs for the current and future climate simulations are shown in the leftmost columns of Tables 1 and 2, respectively. A comparison of these averages for all TCs showed many significant differences ($p < 0.01$). The average TCG locations were farther to the east and north for the future climate simulations (Figure 3b) than for the current climate simulations (Figure 3a). The average sizes and development rates of future TCs were larger than those of current TCs. The increase in average TC sizes at the TCG time was consistent with Khairoutdinov and Emanuel [28], who showed that the storm size of the initial TC vortex increased under the warming experiments. The future increase in TC size was also analyzed in the projection simulations by previous studies [4,29–31] which were reviewed in Knutson et al. [2].

Table 2. Statistical summary of TC characteristics associated with TCG environmental factors from June to November over a 5600-year analysis period in future climate simulations using the d4PDF dataset. Significant differences are indicated in bold and italic fonts (Student's *t*-test; $p < 0.01$); changes are shown below each parameter.

	ALL	SL	CR	GY	EW	PTC	UCF
Longitude at TCG time (°N)	138.5	*137.2*	*141.8*	*133.8*	*140.0*	*141.1*	*144.0*
	+1.1	*+2.0*	*+0.8*	*+0.9*	*−0.6*	*−0.3*	*+5.2*
Latitude at TCG time (°E)	18.1	*18.0*	*18.9*	*17.9*	*19.2*	*15.3*	*19.3*
	+1.5	*+1.4*	*+1.7*	*+1.6*	*+1.5*	*+1.8*	*+1.7*
TC size at TCG time (km)	352.2	*364.8*	*332.7*	*367.2*	*330.7*	*285.6*	*371.8*
	+18.7	*+21.6*	*+11.0*	*+18.0*	*+13.7*	*+23.5*	*+13.2*
Development rate (m s^{-1} 6 h^{-1})	0.91	*0.94*	*0.82*	*0.80*	*0.93*	*0.97*	*0.78*
	+0.30	*+0.31*	*+0.26*	*+0.25*	*+0.33*	*+0.28*	*+0.23*
Movement speed (km h^{-1})	16.1	*15.7*	*16.7*	*15.2*	*17.3*	*16.7*	*16.3*
	+0.1	*+0.3*	*+0.1*	*+0.5*	*−0.2*	*−0.9*	*+0.7*
Duration of development (day)	2.6	*2.7*	*2.7*	*2.4*	*2.5*	*2.8*	*2.3*
	−0.5	*−0.5*	*−0.6*	*−0.5*	*−0.5*	*−0.5*	*−0.5*
LMI (m s^{-1})	29.1	*29.8*	*27.6*	*26.3*	*30.1*	*30.6*	*24.6*
	+3.2	*+3.4*	*+2.2*	*+2.9*	*+4.3*	*+2.8*	*+2.2*

The average movement speed of TCs during the development stage was slightly higher for the future climate simulations than for the current climate simulations. The average duration of the development stage was shorter for future TCs than for current TCs. The average LMI of future TCs (Figure 5) was higher than those of current TCs, which is consistent with the findings of previous studies.

Figure 4 and Table 2 show differences in average TC characteristics stratified by TCG factors between current and future climate simulations. The stratified average characteristics were similar to those of all TCs, except for average TCG location and movement speed during the development stage for EW-TCs and PTC-TCs. Although the average locations of SL-TCs, CR-TCs, GY-TCs, and UCF-TCs were farther east under future climate change, those of EW-TCs and PTC-TCs were farther west.

Therefore, the TCG classified with monsoon related factors tend to shift eastward, and that with easterly related factors tends to shift westward.

Two modes of distribution change in EW-TCs location can be found in the future climate simulations, expanding to the north-eastward broadly and increasing near the northern part of Philippine sea and South China Sea locally. The former trend is similar to monsoon related factors. But the contribution of the TCG increase around Philippine island results in the statistical significance for the westward shift of the EW-TCs genesis location in the future climate simulations.

Average movement speeds during the development stage of EW-TCs and PTC-TCs were lower in the future climate simulations than those of the current climate simulations. Yamaguchi et al. [32] found that the annual average of TC movement speed under future climate change simulated using d4PDF data had variation among latitudes. Since the dependency of the TCG location to the TCG factors was found in this study, the TC movement speed may have been dependent on TCG factors.

Most TC characteristics among the different TCG factors in the future climate simulations (Table 2) showed the similar for current climate simulations (Table 1), with some exceptions. Interestingly, the average LMI (Figure 5) and TC development rate were significantly higher for EW-TCs than for other TCG factors except PTC in the future climate simulations, but not in the current climate simulations. The average TC development rates for all factors were higher in the future climate simulations; however, the largest increase in the average development rates for EW-TCs resulted in the largest increase in LMI for EW-TC (Table 2).

3.3. Future Changes in Environmental Physical Parameters

This study further analyzed the changes in SST and atmospheric parameters associated with TCs from June to November between current and future climate simulations. The average TC characteristics at the both time of TCG and maturity under current and future simulated climate simulations are shown in Tables 3 and 4, respectively. Future SST averages for all TCs and each TCG factor were about 3.0 K higher than those in the current climate simulations at both the time of TCG and maturity.

Table 3. Statistical summary of sea surface temperatures (SSTs) and atmospheric parameters near TCs associated with TCG environmental factors from June to November over the 6000-year analysis period in current climate simulations performed using the d4PDF dataset. Significant differences are indicated in bold and italic fonts (Student's *t*-test; $p < 0.01$). SHR, difference in horizontal wind speed between 200 and 850 hPa; HMD, specific humidity at 500 hPa; VOR850, relative vorticity at 850 hPa.

	ALL	SL	CR	GY	EW	PTC	UCF
SST at TCG time (K)	301.9	301.9	*302.1*	*301.8*	*301.8*	*301.7*	*301.8*
SHR at TCG time (m s^{-1})	13.5	*13.7*	13.5	13.5	*13.3*	13.5	13.6
HMD at TCG time (kg kg^{-1})	4.14	*4.18*	*4.25*	4.17	3.86	*4.04*	*4.09*
VOR850 at TCG time ($\times 10^{-5}$ s^{-1})	2.71	*2.83*	*2.56*	*2.60*	*2.49*	*2.86*	*2.45*
SST at mature time (K)	301.0	301.0	301.0	301.1	*300.8*	*300.7*	301.0
SHR at mature time (m s^{-1})	17.9	*18.1*	17.9	*17.2*	17.6	18.0	*17.1*
HMD at mature time (kg kg^{-1})	3.73	*3.78*	*3.83*	*3.85*	*3.44*	*3.44*	3.72
VOR850 at mature time ($\times 10^{-5}$ s^{-1})	3.34	*3.43*	*3.40*	*3.14*	*2.99*	*3.47*	*3.05*

Table 4. Statistical summary of SSTs and atmospheric parameters near TCs associated with TCG environmental factors from June to November over the 5600-year analysis period in future climate simulations performed using the d4PDF dataset. Significant differences are indicated in bold and italic fonts (Student's *t*-test; $p < 0.01$); changes are shown below each parameter.

	ALL	SL	CR	GY	EW	PTC	UCF
SST at TCG time (K)	304.9	304.9	*305.0*	*304.8*	*304.7*	*304.8*	*304.8*
	+3.0	+3.0	+2.9	+3.0	+2.9	+3.0	+3.0

Table 4. *Cont.*

	ALL	SL	CR	GY	EW	PTC	UCF
SHR at TCG time (m s^{-1})	14.1	14.2	*13.7*	14.2	*13.9*	*14.4*	*14.4*
	+0.6	*+0.5*	*+0.5*	*+0.8*	*+0.5*	*+0.5*	*+0.8*
HMD at TCG time (kg kg^{-1})	5.81	*5.87*	*5.95*	*5.78*	*5.47*	*5.69*	*5.65*
	+1.66	*+1.69*	*+1.70*	*+1.62*	*+1.61*	*+1.66*	*+1.56*
VOR850 at TCG time ($\times 10^{-5}$ s^{-1})	3.10	*3.23*	*2.90*	*2.92*	*2.96*	*3.21*	*2.78*
	+0.39	*+0.40*	*+0.34*	*+0.32*	*+0.47*	*+0.35*	*+0.33*
SST at mature time(K)	304.0	304.1	304.0	304.2	303.9	304.0	304.1
	+3.1	*+3.1*	*+3.0*	*+3.1*	*+3.1*	*+3.2*	*+3.0*
SHR at mature time (m s^{-1})	18.1	*18.3*	*17.9*	*17.7*	*18.2*	*18.4*	*17.3*
	+0.3	*+0.2*	*+0.0*	*+0.5*	*+0.6*	*+0.2*	*+0.2*
HMD at mature time (kg kg^{-1})	5.27	*5.36*	*5.45*	*5.34*	*4.86*	*4.89*	*5.31*
	+1.54	*+1.57*	*+1.61*	*+1.49*	*+1.42*	*+1.45*	*+1.50*
VOR850 at mature time ($\times 10^{-5}$ s^{-1})	3.76	*3.89*	*3.73*	*3.48*	*3.47*	*3.84*	*3.28*
	+0.42	*+0.46*	*+0.33*	*+0.34*	*+0.48*	*+0.37*	*+0.23*

The atmospheric parameters SHR, HMD, and VOR850 were also significantly higher for all TCs in the future climate simulations than in the current climate simulations ($p < 0.01$). Figure 6 shows the differences between current and future climate simulations for the 850-hPa horizontal winds at the time of TCG in summer and autumn. A trend of anomalous westerlies appeared in an area from 10–20 °N of the central Pacific, which enhances cyclonic circulation anomalies over the northeastern part of the WNP in the future climate simulation. The cyclonic circulation anomalies were favorable for TCG in this region, resulting in the northeast shift of the average TCG locations in the future climate simulation.

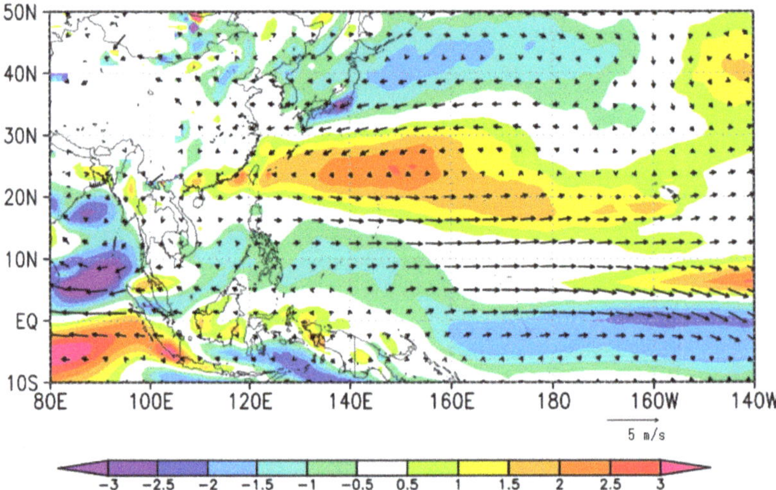

Figure 6. Future changes in horizontal winds (m s^{-1} arrows, scale at bottom) and relative vorticity ($\times 10^{-6}$ s^{-1} color) at 850 hPa at the time of TCG for TCs from June to November over the analysis period in current and future climate simulations performed using the d4PDF.

The average SSTs near CR-TCs (GY-TCs, EW-TCs, PTC-TCs, and UCF-TCs) at the TCG time were higher (lower) than those near TCs associated with other factors in the current climate simulations. At mature time, the average SSTs near EW-TCs and PTC-TCs were lower than those near TCs associated with the other factors. Characteristics of SSTs among the TCG factors in the future climate simulations were similar to those in the current climate simulations.

In the current climate simulations, the average SHR near SL-TCs (EW-TCs) was higher (lower) at the TCG time, whereas the SHR near SL-TCs (GY-TCs, EW-TCs, and UCF-TCs) was higher (lower) at the mature time. In the future climate simulations, the average SHR around PTC-TCs and UCF-TCs (CR-TCs and EW-TCs) was higher (lower) at the TCG time, whereas that around SL-TCs and PTC-TCs (CR-TCs, GY-TCs, and UCF-TCs) was higher (lower) at mature time.

At both time of TCG and maturity in the current climate simulations, average HMD was higher around SL-TCs, CR-TCs, and GY-TCs, due to the Asian monsoon. In contrast, drier conditions were observed around EW-TCs, PTC-TCs, and UCF-TCs. Characteristics of the average HMD among the TCG factors in the future climate simulations were similar to those in the current climate simulations.

At both time of TCG and maturity in the current climate simulations, the average VOR850 was higher near SL-TCs and PTC-TCs, and lower near GY-TCs, EW-TCs, and UCF-TCs ($p < 0.01$). Characteristics of the VOR850 among the TCG factors in the future climate simulations were similar to those in the current climate simulations. Differences in the averages of physical environmental parameters for the different TCG factors between current and future climate simulations are shown in Table 4. Interestingly, changes in VOR850 of each factor were similar to changes in development rate at the TCG time (Table 2). Figure 7 shows the relationship between future changes in VOR850 and development rate stratified by TCG factors. The increases in the average development rate were correlated with the increases in VOR850 (correlation coefficient = 0.91). Thus, higher average development rate of future EW-TCs than those of other factors was also caused by this highest increase in VOR850 around EW-TCs, supported by higher SSTs and wetter conditions under the future climate simulations.

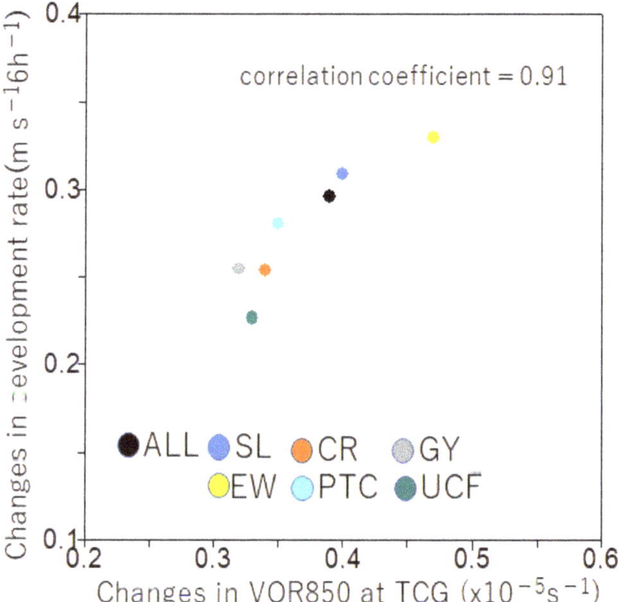

Figure 7. Scatterplot of the relationship between changes in VOR850 and development rate between current and future climate simulations stratified by TCG factors.

4. Discussion

This study evaluated the differences in average TC characteristics over the WNP between current and future climate simulations. The average TCG locations were farther to the east and to the north in the future climate simulations than in the current climate simulations. Murakami et al. [33] and

Yokoi et al. [34] showed an eastward shift in the positions of northward-recurving TC tracks and TCG location over the WNP under future climate. They discussed that the eastward shift was caused by changes in dynamical environmental conditions such as lower-tropospheric relative vorticity, directly resulted from the eastward extension of the monsoon trough. By using simulations from CMIP5, Wang and Wu [35] showed the monsoon trough over the WNP was intensify and extend eastward in the warming scenarios. We have extended their work to examine the contributions of an eastward shift in the TCG location stratified by the environmental factors that influenced TCG. In the future climate simulations, the average TCG locations of SL-TCs, CR-TCs, and GY-TCs, which are related to Asian monsoon, were farther east than those of other factors. Yokoi and Takayabu [26] examined the global warming impact on TCG over the WNP projected by five models that participate in Phase 3 of the Coupled Model Intercomparison Project (CMIP3). All of the five models projected an increasing trend of the TCG frequency over the central North Pacific. They concluded that the increasing trend over the central North Pacific resulted from changes of SST and large-scale circulation field over the central North Pacific that exhibit an El Niño-like pattern. The El Niño-like pattern was characterized by the relative vorticity in the lower troposphere over the central North Pacific that became more favorable for TCG. Their finding was consistent with the results of this study. We found that the average LMI (Figure 5) and TC development rate (Table 2) in the future climate simulations were significantly higher for EW-TCs, mainly occurred over the central North Pacific. In the future climate, the higher development rate of EW-TCs than those of other factors was caused by SST and enhanced cyclonic circulation anomalies (Figure 6) around TCG region of EW-TCs.

The large-scale flow pattern over the WNP changed in the future climate simulations (Figure 6), consistent with previous studies [36,37]. However, the contributions of TCG stratified by environmental factors did not so much change in the future climate simulations. Since the contributions of TCG environment associated with Asian monsoon for the TCG frequency over the WNP, namely SL and CR, were large (Figure 1), the future change of TCG frequency depends on that of monsoon activity. On the other hand, the increases in average LMI projected for all future TCs over the WNP, consistent with several previous studies, were caused by large contributions from the average LMI of future EW-TCs. Our results suggest that the frequency of EW-TCs in the warming scenarios is important for TC projection research.

5. Conclusions

This study investigated future changes in the frequency of TCG and TC characteristics in summer and autumn over the WNP, stratified by TCG environmental factors classified by Ritchie and Holland [7], using the d4PDF large-scale ensemble dataset and the TGS method. The TCG factors explained the environmental conditions of TCG under the future climate simulations, as well as under current climate simulations. Among summer and autumn TCs, SL-TCs (54.4% of all TCs) were about 3.8% more frequent in the future climate simulations than in the current climate simulations, whereas CR-TCs (15.0%) were about 4.4% less frequent and the frequency of EW-TCs (11.9%) and PTC-TCs (7.3%) changed by <1%, respectively. The frequencies of GY-TCs (7.6%) and UCF-TCs (3.7%) were unchanged between the climate simulations.

The statistical characteristics of TCs stratified by TCG factors were similar between the current and future climate simulations, except that average LMI of EW-TCs was significantly higher than those of other TCG factors in the future climate simulations, due to the higher development rate, higher SSTs, higher VOR850, and wetter conditions associated with EW in the future climate simulations. Together, these results suggest that the high increases in average LMI projected for all future TCs over the WNP in several previous studies are due to the relatively high contribution of EW-TCs. Our elucidation of future trends in EW-TCs will be important for disaster prevention under climate change.

Author Contributions: Conceptualization, H.F.; methodology, H.F., K.Y., and R.Y.; formal analysis, H.F.; writing—original draft preparation, H.F.; writing—review and editing, H.F., K.Y., and R.Y. All authors have read and agreed to the published version of the manuscript.

Funding: This research was funded by Ministry of Education, Culture, Sports, Science and Technology (MEXT), Japan, KAKENHI Grants JP19H05696, and partly supported by KAKENHI Grants JP19H00705 and the Integrated Research Program for Advancing Climate Models (TOUGOU) Grant JPMXD0717935498 from MEXT, Japan.

Acknowledgments: The authors are grateful to the editor and anonymous reviewers for their useful and critical comments. This study used d4PDF produced with the Earth Simulator jointly by science programs (SOUSEI, TOUGOU, SI-CAT, DIAS) of the Ministry of Education, Culture, Sports, Science, and Technology (MEXT), Japan. The d4PDF data set is available from the DIAS website (https://www.diasjp.net/en/). This work is supported by MEXT KAKENHI Grants 17H02956, 17K14398, JP19H05696 and TOUGOU program.

Conflicts of Interest: The authors declare no conflict of interest.

References

1. Yoshida, K.; Sugi, M.; Mizuta, R.; Murakami, H.; Ishii, M. Future Changes in Tropical Cyclone Activity in High-Resolution Large-Ensemble Simulations. *Geophys. Res. Lett.* **2017**, *44*, 9910–9917. [CrossRef]
2. Knutson, T.; Camargo, S.J.; Chan, J.C.L.; Emanuel, K.; Chang-Hoi, H.; Kossin, J.; Mohapatra, M.; Satoh, M.; Sugi, M.; Walsh, K.; et al. Tropical Cyclones and Climate Change Assessment: Part II: Projected Response to Anthropogenic Warming. *Bull. Am. Meteor. Soc.* **2020**, *101*, E303–E322. [CrossRef]
3. Roberts, M.J.; Vidale, P.L.; Mizielinski, M.S.; Demory, M.-E.; Schiemann, R.; Strachan, J.; Hodges, K.; Bell, R.; Camp, J. Tropical Cyclones in the UPSCALE Ensemble of High-Resolution Global Climate Models. *J. Clim.* **2015**, *28*, 574–596. [CrossRef]
4. Knutson, T.R.; Sirutis, J.J.; Zhao, M.; Tuleya, R.E.; Bender, M.; Vecchi, G.A.; Villarini, G.; Chavas, D. Global Projections of Intense Tropical Cyclone Activity for the Late Twenty-First Century from Dynamical Downscaling of CMIP5/RCP4.5 Scenarios. *J. Clim.* **2015**, *28*, 7203–7224. [CrossRef]
5. Bhatia, K.; Vecchi, G.; Murakami, H.; Underwood, S.; Kossin, J. Projected Response of Tropical Cyclone Intensity and Intensification in a Global Climate Model. *J. Clim.* **2018**, *31*, 8281–8303. [CrossRef]
6. Zhang, L.; Karnauskas, K.B.; Donnelly, J.P.; Emanuel, K. Response of the North Pacific Tropical Cyclone Climatology to Global Warming: Application of Dynamical Downscaling to CMIP5 Models. *J. Clim.* **2017**, *30*, 1233–1243. [CrossRef]
7. Ritchie, E.A.; Holland, G.J. Large-scale patterns associated with tropical cyclogenesis in the western Pacific. *Mon. Weather Rev.* **1999**, *127*, 2027–2043. [CrossRef]
8. Lander, M.A. Description of a monsoon gyre and its effects on the tropical cyclones in the western North Pacific during August 1991. *Weather Forecast.* **1994**, *9*, 640–654. [CrossRef]
9. Chen, S.S.; Houze, R.A., Jr.; Mapes, B.E. Multiscale variability of deep convection in relation to large-scale circulation in TOGA COARE. *J. Atmos. Sci.* **1996**, *53*, 1380–1409. [CrossRef]
10. Heta, Y. An analysis of tropical wind fields in relation to typhoon formation over the western Pacific. *J. Meteor. Soc. Jpn.* **1990**, *68*, 65–77. [CrossRef]
11. Heta, Y. The origin of tropical disturbances in the equatorial Pacific. *J. Meteorol. Soc. Jpn.* **1991**, *69*, 337–351. [CrossRef]
12. McDonald, N.R. The decay of cyclonic eddies by Rossby wave radiation. *J. Fluid Mech.* **1998**, *361*, 237–252. [CrossRef]
13. Li, T.; Fu, B. Tropical cyclogenesis associated with Rossby wave energy dispersion of a preexisting typhoon. Part I: Satellite data analyses. *J. Atmos. Sci.* **2006**, *63*, 1377–1389. [CrossRef]
14. Li, T.; Ge, X.; Wang, B.; Zhu, Y. Tropical cyclogenesis associated with Rossby wave energy dispersion of a preexisting typhoon. Part II: Numerical simulations. *J. Atmos. Sci.* **2006**, *63*, 1390–1409. [CrossRef]
15. Yoshida, R.; Ishikawa, H. Environmental factors contributing to tropical cyclone genesis over the Western north Pacific. *Mon. Weather Rev.* **2013**, *141*, 451–467. [CrossRef]
16. Onogi, K. The JRA-25 Reanalysis. *J. Meteorol. Soc. Jpn.* **2007**, *85*, 369–432. [CrossRef]
17. Fudeyasu, H.; Yoshida, R. Western North Pacific Tropical Cyclone Characteristics Stratified by Genesis Environment. *Mon. Weather Rev.* **2018**, *146*, 435–446. [CrossRef]
18. Yamada, Y.; Oouchi, K.; Satoh, M.; Tomita, H.; Yanase, W. Projection of changes in tropical cyclone activity and cloud height due to greenhouse warming: Global cloud-system-resolving approach. *Geophys. Res. Lett.* **2010**, *37*. [CrossRef]

19. Camargo, S.; Tippett, J.; Michael, K.; Sobel, A.; Vecchi, H.; Gabriel, A.; Zhao, M. Testing the performance of tropical cyclone genesis indices in future climates using the HiRAM model. *J. Clim.* **2014**, *27*, 9171–9196. [CrossRef]

20. Mizuta, R.; Yoshimura, H.; Murakami, H.; Matsueda, M.; Endo, H.; Ose, T.; Makiguchi, M.; Hosaka, M.; Sugi, S.; Yukimoto, S.; et al. Climate simulations using MRI-AGCM3.2 with 20-km Grid. *J. Meteorol. Soc. Jpn.* **2012**, *90A*, 233–258. [CrossRef]

21. Hirahara, S.; Ishii, M.; Fukuda, Y. Centennial-scale sea surface temperature analysis and its uncertainty. *J. Clim.* **2014**, *27*, 57–75. [CrossRef]

22. Mizuta, R.; Murata, A.; Ishii, M.; Shiogama, H.; Hibino, K.; Mori, N.; Arakawa, O.; Imada, Y.; Yoshida, K.; Aoyagi, T.; et al. Over 5000 years of ensemble future climate simulations by 60 km global and 20 km regional atmospheric models. *Bull. Amer. Meteorol. Soc.* **2017**, *98*, 1383–1398. [CrossRef]

23. Murakami, H.; Mizuta, R.; Shindo, E. Future changes in tropical cyclone activity projected by multi-physics and multi-SST ensemble experiments using the 60-km-mesh MRI-AGCM. *Clim. Dyn.* **2012**, *39*, 2569–2584. [CrossRef]

24. Fudeyasu, H.; Yoshida, R. Statistical Analysis of the Relationship between Upper Tropospheric Cold Lows and Tropical Cyclone Genesis over the Western North Pacific. *J. Meteorol. Soc. Jpn.* **2019**, *97*, 439–451. [CrossRef]

25. Knapp, K.R.; Kruk, M.C.; Levinson, D.H.; Diamond, H.J.; Neumann, C.J. The International Best Track Archive for Climate Stewardship (IBTrACS). *Bull. Am. Meteorol. Soc.* **2010**, *91*, 363–376. [CrossRef]

26. Yokoi, S.; Takayabu, Y.N. Multi-model projection of global warming impact on tropical cyclone genesis frequency over the Western North Pacific. *J. Meteorol. Soc. Jpn.* **2009**, *87*, 525–538. [CrossRef]

27. Sugi, M.; Yamada, Y.; Yoshida, K.; Mizuta, R.; Nakano, M.; Kodama, C.; Satoh, M. Future changes in the global frequency of tropical cyclones seeds. *SOLA* **2020**, *16*, 70–74. [CrossRef]

28. Khairoutdinov, M.F.; Emanuel, K. Rotating radiative-convective equilibrium sismulated by a cloud-resolving model. *J. Adv. Model. Earth Sys.* **2013**, *5*, 816–825. [CrossRef]

29. Kim, H.-S.; Vecchi, G.A.; Knutson, T.R.; Anderson, W.G.; Delworth, T.L.; Rosati, A.; Zeng, F.; Zhao, M. 2014: Tropical cyclone simulation and response to CO_2 doubling in the GFDL CM2.5 high-resolution coupled climate model. *J. Clim.* **2014**, *27*, 8034–8054. [CrossRef]

30. Yamada, Y.; Satoh, M.; Sugi, M.; Kodama, C.; Noda, A.T.; Nakano, M.; Nasuno, T. Response of tropical cyclone activity and structure to global warming in a high-resolution global nonhydrostatic model. *J. Clim.* **2017**, *30*, 9703–9724. [CrossRef]

31. Gutmann, E.D.; Rasmussen, R.M.; Liu, C.; Ikeda, K.; Bruyere, C.L.; Done, J.; Garre, L.; Friis-Hansen, P.; Veldore, V. Changes in hurricanes from a 13-yr convection-permitting pseudo-global warming simulation. *J. Clim.* **2018**, *31*, 3643–3657. [CrossRef]

32. Yamaguchi, M.; Chan, J.C.; Yoshida, K.; Mizuta, R. Global warming changes tropical cyclone translation speed. *Nat. Commun.* **2020**, *11*, 1–7. [CrossRef] [PubMed]

33. Murakami, H.; Wang, B.; Kitoh, A. Future change of western North Pacific typhoons: Projections by a 20-km-mesh global atmospheric model. *J. Clim.* **2011**, *24*, 1154–1169. [CrossRef]

34. Yokoi, S.; Takahashi, C.; Yasunaga, K.; Shirooka, R. Multi-model projection of tropical cyclone genesis frequency over the Western North Pacific: CMIP5 results. *SOLA* **2012**, *8*, 137–140. [CrossRef]

35. Wang, C.; Wu, L. Future changes of the monsoon trough: Sensitivity to sea surface temperature gradient and implications for tropical cyclone activity. *Earth's Future* **2018**, *6*, 919–936. [CrossRef]

36. Li, X.; Ting, M.; Li, C.; Henderson, N. Mechanisms of Asian Summer Monsoon Changes in Response to Anthropogenic Forcing in CMIP5 Models. *J. Clim.* **2015**, *28*, 4107–4125. [CrossRef]

37. Zhang, L.; Li, T. Relative roles of anthropogenic aerosols and greenhouse gases in land and oceanic monsoon changes during past 156 years in CMIP5 models. *Geophys. Res. Lett.* **2016**, *43*, 5295–5301. [CrossRef]

Publisher's Note: MDPI stays neutral with regard to jurisdictional claims in published maps and institutional affiliations.

Article

Future Changes in Tropical Cyclone and Easterly Wave Characteristics over Tropical North America

Christian Dominguez [1,*], James M. Done [2] and Cindy L. Bruyère [2,3]

[1]   Centro de Ciencias de la Atmósfera, Universidad Nacional Autónoma de México (UNAM), Mexico City 04510, Mexico

[2]   National Center for Atmospheric Research, Boulder, CO 80301, USA; done@ucar.edu (J.M.D.); bruyerec@ucar.edu (C.L.B.)

[3]   Environmental Sciences and Management, North-West University, Potchefstroom 2531, South Africa

*   Correspondence: dosach@atmosfera.unam.mx

Abstract: Tropical Cyclones (TCs) and Easterly Waves (EWs) are the most important phenomena in Tropical North America. Thus, examining their future changes is crucial for adaptation and mitigation strategies. The Community Earth System Model drove a three-member regional model multi-physics ensemble under the Representative Concentration Pathways 8.5 emission scenario for creating four future scenarios (2020–2030, 2030–2040, 2050–2060, 2080–2090). These future climate runs were analyzed to determine changes in EW and TC features: rainfall, track density, contribution to seasonal rainfall, and tropical cyclogenesis. Our study reveals that a mean increase of at least 40% in the mean annual TC precipitation is projected over northern Mexico and southwestern USA. Slight positive changes in EW track density are projected southwards 10° N over the North Atlantic Ocean for the 2050–2060 and 2080–2090 periods. Over the Eastern Pacific Ocean, a mean increment in the EW activity is projected westwards across the future decades. Furthermore, a mean reduction by up to 60% of EW rainfall, mainly over the Caribbean region, Gulf of Mexico, and central-southern Mexico, is projected for the future decades. Tropical cyclogenesis over both basins slightly changes in future scenarios (not significant). We concluded that these variations could have significant impacts on regional precipitation.

Keywords: tropical cyclones; easterly waves; future changes in rainfall; changes in tropical cyclogenesis

Citation: Dominguez, C.; Done, J.M.; Bruyère, C.L. Future Changes in Tropical Cyclone and Easterly Wave Characteristics over Tropical North America. *Oceans* **2021**, *2*, 429–447. https://doi.org/10.3390/oceans2020024

Academic Editor: Hiroyuki Murakami

Received: 14 April 2021
Accepted: 7 June 2021
Published: 10 June 2021

Publisher's Note: MDPI stays neutral with regard to jurisdictional claims in published maps and institutional affiliations.

1. Introduction

Tropical Cyclones (TCs) are widely known worldwide for their destructive effects on the continental landmass and the high economic losses associated with their passage [1] For instance, TCs are linked with 86.5% of the annual disaster costs in Mexico, where the main hazard is the heavy rainfall they produced [2]. Dominguez et al. [3] show that disaster costs in Mexico remarkably depend on natural variability, such as ENSO, because the national social vulnerability is high. Tropical storms or even tropical depressions can produce such huge amounts of rainfall that one single event can increase the chances of disaster occurrence. However, TCs also play an important role as moisture carriers for arid and semiarid regions, where they can contribute up to 30% of the accumulated seasonal rainfall and help to ameliorate water scarcity [4]. In that sense, exploring future regional changes in TC features (e.g., rainfall and frequency) is considered necessary for water availability projections and domestic decision makers.

Future scenarios for tropical cyclone activity reveal different results that depend on the basin, model configuration, representation of ocean-air feedback, periods of study, and the Representative Concentration Pathways (RCPs) that were considered, among others [5–8]. For example, Bacmeister et al. [9] and Wehner et al. [10] explored high-resolution simulations from the Community Atmosphere Model and found that while high-resolution models do not largely improve simulated climate, interannual variability of TCs is better simulated. Recently, Bacmeister et al. [11] used the Community Earth System Model (CESM) at a 28 km spatial resolution and determined that TC activity globally decreases in the RCP4.5 and RCP8.5 future scenarios. Apart from that, the High Resolution Model Intercomparison Project (HighResMIP v1.0) was created in 2016 as a coordinated effort among several international centers to investigate the impact of increased horizontal resolution based on multimodel ensembles and is part of the Coupled Model Intercomparison Experiment Phase 6 (CMIP6) [12]. The main goal of HighResMIP v1.0 is to determine the dynamical and physical mechanisms that lead to differences in model simulations, which, in the end, will foster confidence in model skill or help to understand model constraints [12].

As part of HighResMIP v1.0, Bao et al. [13] recently experimented with increasing horizontal resolution from 1.0° to 0.25° on one General Climate Model (GCM), called CAS FGOALS-f3-H, in 2020. It was found that this model produced more long-lived TCs in all basins from 2001–2010, when compared with 1.0° spatial resolution simulations. Additionally, Roberts et al. [14,15] explored the future TC activity in a high-resolution multimodel ensemble from HighResMIP. It is important to mention that these studies are among the first to use high-resolution GCMs to analyze future changes in TCs. They found that an increase in the model's spatial resolution leads to a decrease in negative density bias, mainly over the North Atlantic (NATL) and Eastern Pacific (EPAC) Oceans. These results also show that TC activity over the NATL will decrease, and tracks over EPAC will experience a poleward shift in the future (2020–2050).

Interestingly, Knutson et al. [16] previously found similar results over NATL using a regional climate model (RCM) based on CMIP3 and CMIP5 scenarios: a decrease (~20%) in TC frequency, similar to the results found by Roberts et al. [14], and an increase (~60% on average) in TC intensity. However, the results for future TC activity over EPAC are inconsistent. While GCMs from CMIP5 show a reduction in frequency, RCM outputs point out an increase over this basin [8]. In terms of rainfall, Knutson et al. [17,18] analyzed some studies that explored rainfall-rate change for NATL and EPAC and determined that TC rainfall will increase ~5–25% in both basins by the end of the century. The latter results from an increase in the large-scale atmospheric water vapor content, tropospheric and oceanic warming.

It is worth mentioning that until now, several research questions regarding changes in features of tropical phenomena (i.e., TC precipitation or tropical-wave-related rainfall) have been only addressed by the RCM experiments, focusing on specific basins. Although HighResMIP can provide valuable information about future changes in TCs due to the high spatial resolution, it would be ideal to compare the high-resolution GCM outputs with RCM simulations. In this sense, this comparison will let us understand to what extent RCMs can still produce advantageous projections of changes in tropical phenomena. This information could inform future tasks of the Coordinated Regional Climate Downscaling Experiment, known as CORDEX.

Easterly Waves (EWs), commonly known as tropical-depression-type waves, are Rossby wave-like disturbances over the tropics that can be precursors of TCs [19]. These waves propagate westwards and last up to 6 days over NATL [20] and up to 8 days over EPAC [21]. Over the NATL basin, two waveguides can be distinguished: one to the south of 10° N (hereafter southern waveguide) and the other to the north, around 15° N (hereafter northern waveguide); each playing different roles in tropical cyclogenesis [22]. Overall, it has been found that ~50% and ~70% of TCs from NATL and EPAC originate from EWs, respectively [23–25]. Apart from being important for tropical cyclogenesis, EWs are also relevant moisture carriers for the tropics because they can contribute up to 60% of seasonal accumulated rainfall over Tropical North America (TNA) [25]. However, this role has been less explored than the heavy rainfall produced by TCs.

Moreover, most EW studies have focused on exploring the relationship between EWs and the West Africa monsoon [26,27]. For example, Martin and Thorncroft [28] found that CMIP5 models have problems capturing EW propagation, maintenance, and growth mechanisms over the west African coast and concluded that high-resolution models can better represent EW dynamics. Furthermore, Kebe et al. [29] explained that a projected reduction in the EW activity impacts the seasonal summer precipitation over West Africa for the 2080–2099 period. However, future changes in EW rainfall over the TNA remain unexplored. Consequently, no studies investigate how EW rainfall over TNA will change in the future under climate change conditions. Having this information is crucial for local stakeholders and farmers that will struggle with climate-change effects.

The main objective of this study is to explore future changes in the rainfall produced by TCs and EWs over TNA using a regional climate simulation ensemble under a future climate scenario. Our emphasis is on analyzing the response of the tropical phenomena characteristics to different representations of large-scale forcing rather than investigating the full set of possible future scenarios. In this sense, our three-member ensemble forced by only one GCM does not sample the full range of future possibilities due to different emission scenarios, internal GCM variability, or diverse ocean responses to radiative forcing. This study investigates potential changes in tropical cyclogenesis, TC rainfall, and EW rainfall, as well as their contribution to summer rainfall (defined here as accumulated precipitation from May to November) of the current climate represented by the 1990–2000 decade, along with two near future decades (2020–2030, 2030–2040) and two far future decades (2050–2060, 2080–2090), by using a three-member ensemble. The results of this study can be relevant for regional climate change adaption strategies in most countries from Latin America. We use dynamical downscaling to provide detailed information about TC and EW characteristics that coarse spatial resolution GCMs could miss (e.g., structure, evolution, and their impact on regional climate). An RCM grid spacing of 36 km has been demonstrated to represent adequately tropical synoptic circulations and features [7]. Some studies show that RCMs realistically captured spatial distribution, evolution, generation, and organized convection of EWs [30,31]. The latter indicates that RCMs can provide valuable regional information about TC and EW frequency changes and their associated rainfall.

The next section describes the reanalysis data and observed tracks for TCs over NATL and EPAC. The method used to attribute rainfall to TCs and EWs, regional climate model configuration, and criteria for tracking tropical phenomena in the RCM outputs are also discussed. Section 3 analyzes the future changes in seasonal rainfall produced by EWs and TCs over TNA. Future changes in the rate of TCs that originate from EWs are also presented. Finally, the summary and conclusions are given in Section 4.

2. Data and Methods

2.1. Reanalysis, TC Tracks, and Tropical Rainfall Attribution

ERAInterim (ERAI) is available from 1979 to 2019 and has a T255 spectral resolution (~80 km) on 60 vertical model levels [32]. TC rainfall was defined using the precipitation from this dataset for the 1990–2000 period. TC tracks whose category is a tropical depression, tropical storm, or hurricane over the NATL and EPAC were obtained from the Hurricane Database, also known as HURDAT, for the same period [33]. Following previous studies [4,34,35], TC-related rainfall is defined as the precipitation inside a 500 km radius of influence from the TC center location provided by HURDAT. This approximation has been widely recognized to compute adequately the rainfall associated with TC passages. We compared the TC rainfall obtained from ERAI with the TC rainfall in the RCM outputs for the 1990–2000 period to determine if regional variations are well captured.

2.2. Regional Climate Model Configuration

One of the advantages of using RCMs is that tropical phenomena, such as TCs and EWs, are better resolved in RCMs than in GCMs, due to their relatively coarser resolution. For this study, we use an existing RCM ensemble described by Bruyère et al. [7]. ERAI reanalysis drove the Weather Research and Forecasting model version 3.4 (hereafter

WRF; [36]) for producing the current climate simulations (1990–2000). The RCM ensemble was forced by using the Reynolds optimum interpolated analysis [37]. The experiments were designed at the National Center for Atmospheric Research (NCAR) in 2013 and have been used in various studies in the American continent. The domain of the experiments ranges from Canada to Brazil and from Hawaii to the mid-Atlantic Ocean (Figure 1). The domain size is enough for capturing EWs that enter the eastern boundary, whose lateral boundary conditions are provided by the ERAI reanalysis for the current climate runs and by the CESM for future climate runs. These EWs become TCs over the NATL within the RCM domain [7]. Furthermore, the fact that the domain boundaries are far away from TNA's continental landmasses lets the regional climate phenomena evolve freely within the WRF domain, as they are not limited by the boundaries [38].

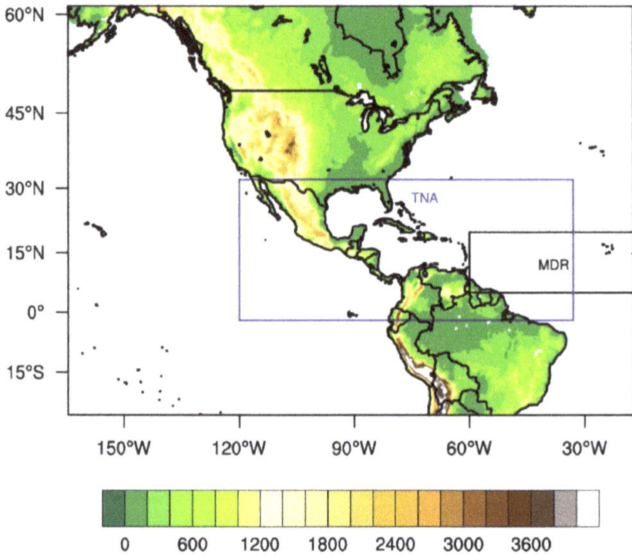

Figure 1. Regional climate model domain. Model terrain height (m) and the two study regions: Tropical North America (TNA) and Main Development Region (MDR).

Three members were chosen from a 24-member model physics ensemble because they performed an adequate simulation of TC activity for the 1990–2000 period, as described in Bruyère et al. [7], and were run at 36 km spatial resolution. The three members have different physical parameterization schemes: radiation scheme (RRTMG [39]); cumulus convection (Kain Fritsch [40], New Simplified Arakawa–Schubert [41], and Tiedtke [42]); microphysics (Thompson [43]) and planetary boundary layer (PBL) (Mellor-Yamada-Janjić [44], and Yonsei University [45]). The Noah scheme [46] was the only land surface scheme used for the experiments. The three members are listed in Table 1. The four letters assigned to the three members represent a combination of parameterizations. For example, the member rtty means that RRTMG is the radiation scheme, Tiedtke is the cumulus parameterization, Thompson is the microphysics scheme, and the fourth letter represents the Mellor–Yamada–Janjić planetary boundary layer scheme.

Table 1. The physics combinations of the three members used for the future runs. These members use the same radiation (RRTMG) and microphysics (Thompson) scheme. All simulations are conducted at a 36 km grid spacing.

Cumulus Convection	Planet Boundary Layer	
	Mellor–Yamada–Janjić (M)	*Yonsei University (Y)*
Kain Fritsch (K)	RKTM	
New Simplified Arakawa–Schubert (N)		RNTY
Tiedtke (T)		RTTY

In this way, we provide robustness of the physical processes to different representations of atmospheric processes. Future climate runs are designed to provide enough samples of different atmospheric representations. The CESM [47], which is part of the CMIP5 [48], drove the future runs under the RCP8.5 future emission scenario [49]. Biases from this global model were removed before forcing the WRF model, following Bruyère et al. [50].

2.3. Criteria for Tracking Tropical Cyclones and Easterly Waves in Model Outputs

2.3.1. Criteria for Tropical Cyclones and Attribution of Rainfall

TCs were tracked using the algorithm created by Hodges [51] in the current and future climate runs. The TRACK technique follows vorticity centers at 700 mb and only retained vortexes that have these TC characteristics:

1. The sum of the horizontal temperature difference between the storm and its surroundings at 700 mb, 500 mb, and 300 mb is greater than 2 K.
2. The mean 850 mb wind speed is greater than the mean 300 mb wind speed.
3. The 300 mb horizontal temperature difference between the storm and its surroundings is greater than that at 850 mb.
4. The genesis location must be south of $40°$ N.
5. The vorticity centers should retain tropical storm wind intensity for a minimum of 36 h.
6. A threshold of wind speed at 10 m should be met: 12 m/s was for all KF simulations and 10 m/s for all other simulations [7].

Specifically, the wind criteria were created following the wind profile method proposed by Walsh (2007) [52]. These six criteria have been demonstrated to adequately work in WRF simulations at 36 km resolution [38]. The TC-related rainfall was also defined as the 500 km radius of influence from the TC-like vortex center location.

2.3.2. Criteria for Easterly Waves and Attribution of Rainfall

EWs were also tracked at 700 mb using the Hodges algorithm, but with different criteria:

1. Westward movement;
2. A lifetime of at least 2 days;
3. The first detected points are over the oceanic domain 5–$20°$ N;
4. Vorticity centers are 1.8×10^{-5} s^{-1} or greater;
5. Systems travel at least 1000 km;
6. Wind speed at 10 m should be less than 12 m/s for all KF simulations and should be less than 10 m/s for all other simulations.

These objective criteria have been widely used in several studies [20,25,53] and have proved to detect EW frequency adequately. Thus, TCs and EWs are both tracked using the described objective algorithm. EWs and TCs were matched by determining points where EW characteristics are accomplished, and then EWs evolve to meet TC features during the track. In that way, the portion of TCs that come from EWs was obtained for each member.

The EW-related rainfall was defined following previous studies [23,25], where a $15° \times 15°$ region around the center provided by the TRACK technique is used to define precipitation produced by EWs. It is worth mentioning that we suppose that precipitation matches with the positive phase of EWs and that the precipitation is only constrained to the $15° \times 15°$ box. Our approach for considering only the positive phase (represented by positive vorticity) may result in underestimating the rainfall because the negative phase of EWs can still produce shallow convection [27]. This assumption may represent a limitation, and it is discussed in Section 4.

2.3.3. The Kolmogorov–Smirnov Two-Sample Test

We used the Kolmogorov–Smirnov two-sample test (KST) to evaluate if the current and future climate results belong to the same distribution (null hypothesis) [54]. If the null hypothesis cannot be rejected, the current and future runs belong to the same distribution, showing no future shifts in climate. In other words, we assume that radiative forcing of $8.5 \, W/m^2$ associated with an increase in Greenhouse Gas concentration will not uniformly cause changes in future climate for all regions from TNA. All future changes are evaluated at a 95% significance level.

3. Results

3.1. Future Changes in Sea Surface Temperatures and Tropical Precipitation

The CESM provided the sea surface temperatures (SSTs) used to force the three-member ensemble for the four future decades. Mean SST anomalies are defined as the annual mean SST values for four future decades minus the SSTs obtained from the Reynolds optimum interpolated analysis during the 1990–2000 period (Figure 2). The mean SST anomalies are generally projected to increase across the future decades. In the two near-future decades, SST anomalies are shown to rise up to 1.2 K. By the 2050–2060 period, a tropical SST warming is projected to strengthen up to 1.8 K in most regions over the TNA. Notably, the most extreme changes in SSTs will occur in the Gulf of Mexico, part of the Caribbean region, and the belt of 5–15° N over EPAC, where SST anomalies vary from 2.4 K to 3 K, for the 2080–2090 period, under the RCP8.5 future emission scenario (Figure 2). However, this uniform increment in SSTs shown in Figure 2 does not mean a uniform increase in tropical convection activity in general [55].

Apart from that, several studies that only focus on TCs have revealed an increase in TC intensity [18,56,57] and TC-related rainfall [16,58] due to changes in SSTs. In particular, Knutson et al. [18] summarized the key findings of global research about the influence of climate change on TCs. They mentioned that changes in tropical tropospheric temperature and SSTs could influence TC activity. Interestingly, TC rainfall rates were determined to increase globally by about 14% for a 2 °C warming at medium-to-high confidence, being slightly lower for individual basins.

Bruyère et al. [7] estimate that the three members have adequate skill to simulate precipitation and temperature over Mexico and Central America. However, these members also have a dry/wet bias (± 6 mm/day) and a cold/hot bias (± 4 °C/day) for the June-July-August months when compared to the Climate Forecast System Reanalysis (CFSR) during the 1990–2000 period [59]. The member rnty was the best one at simulating the rainfall over TNA. Bruyère et al. [7] also show that tropical rainfall is projected to decrease in regions where the average maximum temperature will increase (i.e., central-southern Mexico and Central America) for the June-July-August months during the far future decades (2050–2060 and 2080–2090). On the contrary, their results indicate a rise in tropical rainfall in some parts of the tropical Pacific Ocean for the four future decades.

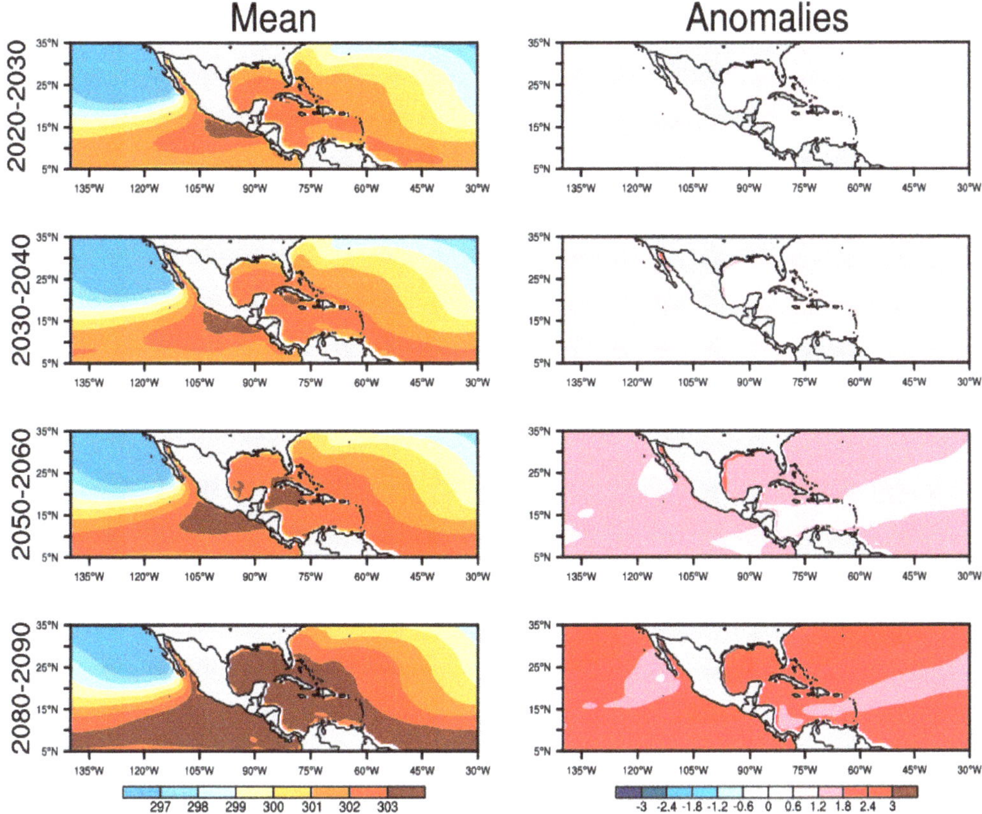

Figure 2. Annual mean (K degrees) and mean anomalies (K degrees) of sea surface temperatures over Tropical North America with respect to the reference period (1990–2000).

In terms of percent change, the three members agree with an increase by up to 50% in the summer rainfall over northern Mexico across the four future decades at a 95% significance level (Figure 3). On the other hand, a decrease of up to 40% in summer rainfall is projected for Central America, northern South America, and the Caribbean region, the changes being more uncertain for the latter region (Figure 3). These results are also consistent with Durán-Quesada et al. [60], who used an ensemble of four GCMs under the RCP2.6, RCP7.0, and RCP8.5 future emission scenarios. Our results suggest that Central America, the Caribbean region, and northern South America will experience droughts in the future, causing water scarcity and impacts in tropical ecosystems [61]. Apart from that, it is notable that non-significant regions increase their area over Central America during the two far future decades, pointing out that uncertainty grows for distant futures.

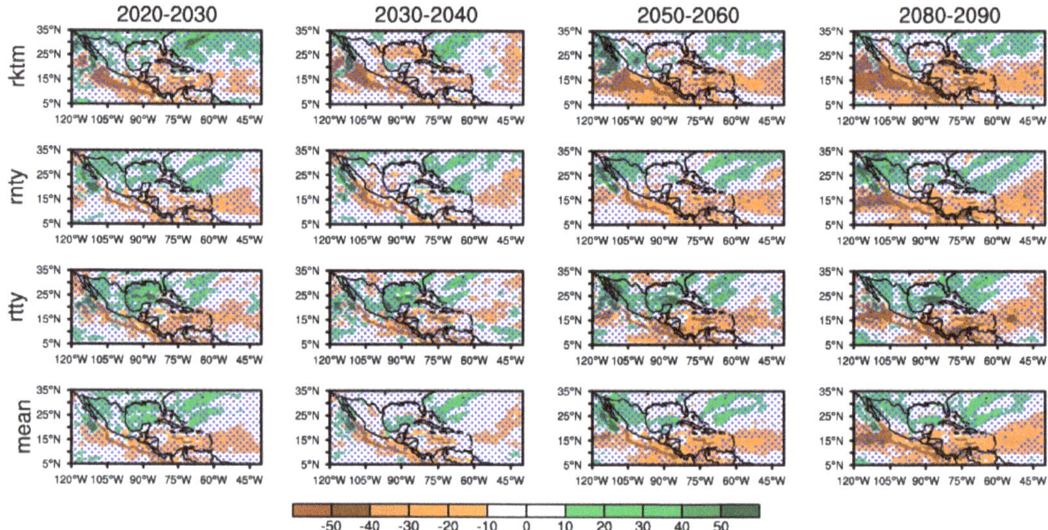

Figure 3. Mean annual percent change (%) in accumulated precipitation from May to November, defined as summer rainfall, with respect to the reference period (1990–2000). The dotted regions indicate a 95% significance level applying the Kolmogorov–Smirnov two-sample test.

3.2. Changes in Tropical Cyclones Features: Rainfall and Track Density

Bruyère et al. [7] show that the cumulus and boundary layer parameterizations have the largest impacts on the annual TC frequency over NATL and EPAC. The observed record of TC activity was 10 ± 3 TCs per year over NATL and 16 ± 4 TCs per year over EPAC. The physical combination of Kain–Fritsch with Mellor–Yamada–Janjić "rktm" registered 12 (17) TCs per year over NATL (EPAC). The combination of Tiedtke with Yonsei University "rtty" produced 8 (14) TCs per year over NATL (EPAC). However, the combination New Simplified Arakawa Schubert/Yonsei University "rnty" generated 9 (6) TCs per year over NATL (EPAC), showing fewer TCs than the observed number over EPAC. In general, TC variability over NATL was successfully captured by the three members during the current climate runs (1990–2000). For the EPAC, only rktm and rtty were able to simulate TC numbers inside the range of one standard deviation.

Mean annual TC track density was computed as the number of tracks per unit area (1° × 1°), and per year from May to November for the three ensemble members (Figure 4a,c,e), the ensemble mean (Figure 4g), and HURDAT tracks (Figure 4i) for the current climate runs. In general, TC track density is successfully represented by the three ensemble members over NATL. However, the spatial distribution of TC tracks is overestimated over EPAC in the belt of 10–15° N by the members rtkm and rtty (Figure 4a,e) when compared to HURDAT (Figure 4i). On the contrary, the member rnty underestimates the mean annual TC track density (Figure 4c), as shown by Bruyère et al. [7]. Interestingly, the ensemble mean adequately represents the observed TC track density over EPAC (Figure 4g), although it slightly overestimates TC density near the Western Central America coast.

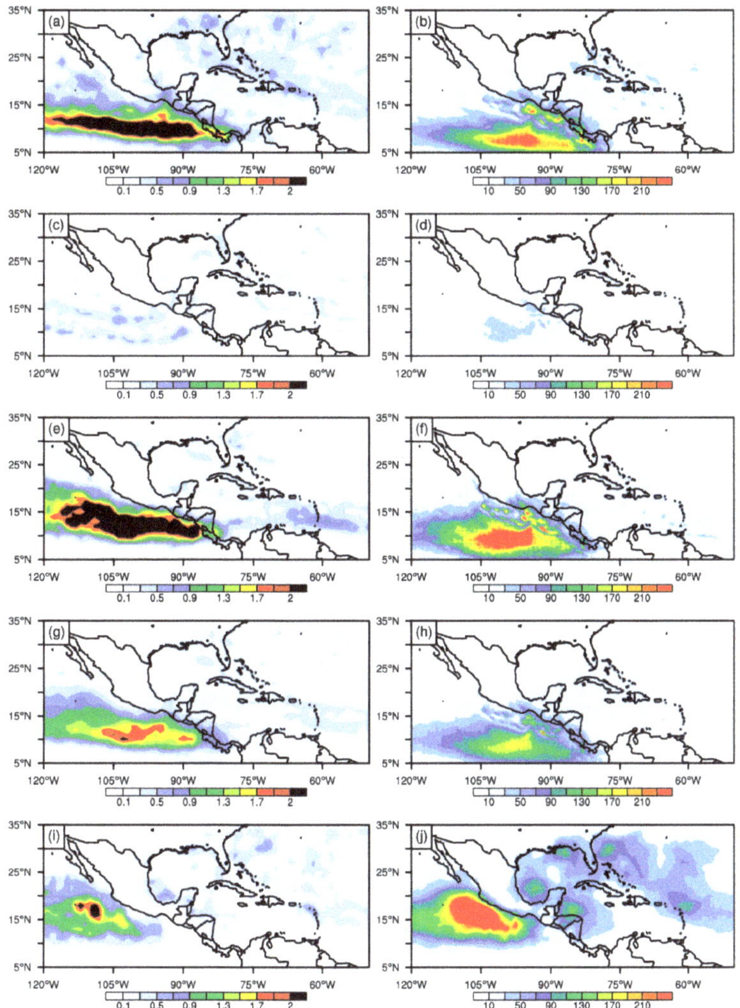

Figure 4. Mean annual tropical cyclone density for three members of the RCM ensemble. (**a**) rktm, (**c**) rnty, (**e**) rtty, (**g**) ensemble mean, (**i**) HURDAT, and mean annual tropical cyclone rainfall for three members of the RCM ensemble: (**h**) rktm, (**d**) rnty, (**f**) rtty, (**h**) ensemble mean, and (**j**) ERAI during the 1990–2000 period.

Mean annual TC-related rainfall was also computed for the 1990–2000 period using the ERAI database. In general, all members acceptably simulate TC precipitation (less than 90 mm/y) over NATL (Figure 4b,d,f). The main difference, when compared with ERAI, is located over the Gulf of Mexico, where TC rainfall is more than 110 mm/yr in some parts (Figure 4j). Over the EPAC, the members rktm and rtty produce their maximum southwards (10–15° N) of the ERAI maximum (Figure 4b,f). This overrepresentation could be related to the number of TCs that are simulated over this belt. Nevertheless, the member rnty poorly represents both TC frequency and TC-related rainfall over EPAC, which balances the ensemble mean, making it similar to observations over this basin.

Jaye et al. [62] reveal that the members rktm and rnty project a decrease in the TC frequency over NATL from the 2020–2030 period through the end of the century. Only the member rtty shows a small increase in TC activity by the 2080–2090 period. Furthermore,

Bruyère et al. [7] show that the ensemble mean also projects a 30% reduction of the TC activity over NATL and a 35% decrease over the Gulf of Mexico, which is partly consistent with Roberts et al. [15], who studied the whole NATL basin. In general, the three members project a reduction in the spatial distribution of TC tracks over the Eastern NATL but show slight spatial differences over the Gulf of Mexico for all the future decades, which are significant at a 95% confidence level (Figure 5). Moreover, the three members and the ensemble mean project significant negative changes in TC activity over EPAC during all the future decades, the most abrupt reduction being in the 105–120° W region and near the Central American coast (where TC activity is overestimated by the members rktm and rtty) by the end of the century (Figure 5). These results are consistent with Torres-Alavez et al. [8], who found a decrease in future TC activity over EPAC using the RegCM4 model driven by three CMIP5 models using RCP2.6 and RCP8.5 for making regional projections over the CORDEX domain.

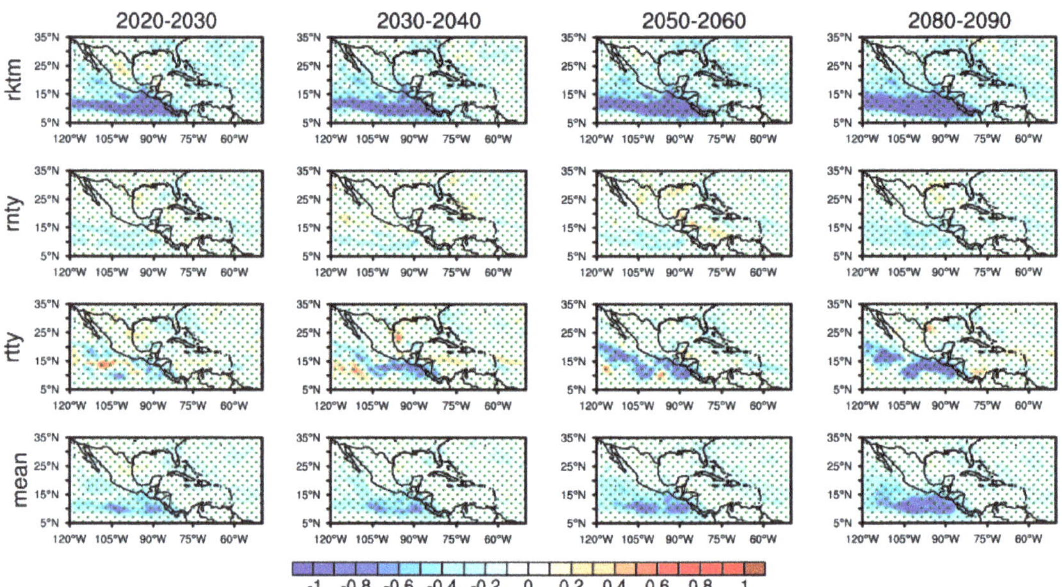

Figure 5. Annual mean anomalies of tropical cyclone density in the future ensemble runs over Tropical North America with respect to the reference period (1990–2000). The dotted regions indicate a 95% significance level applying the Kolmogorov–Smirnov two-sample test.

In terms of TC-related rainfall, Figure 6 shows that all members project a diminishment of up to 80% of the mean TC precipitation over the Caribbean region, southern Mexico, and Central America (not significant) for most decades. This reduction of TC-related rainfall partly explains the decrease in summer rainfall over Central America and the Caribbean region, as shown in Figure 3, as TCs contribute up to 15% of summer rainfall [4]. The RCM outputs have a large uncertainty over the Central American and eastern part of the EPAC, since the KST shows that the samples belong to the same distribution, and this non-significant region mainly expands the area during the last two decades.

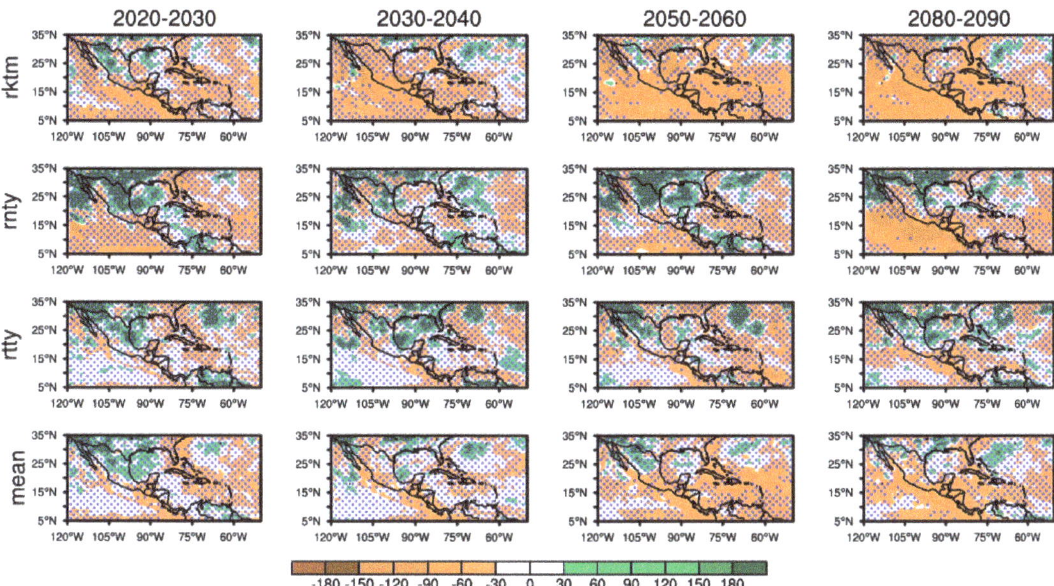

Figure 6. Mean annual percent change (%) in precipitation associated with tropical cyclones in the future ensemble runs with respect to the reference period (1990–2000). The dotted regions indicate a 95% significance level applying the Kolmogorov–Smirnov two-sample test.

A significant increase that ranges from 40% to 200% in mean annual TC precipitation over semiarid regions, as northern Mexico and the southwestern USA, is projected by the members rnty and rtty across the four future decades (Figure 6). The member rktm only projects a rise in mean annual TC precipitation for the 2020–2030 and 2050–2060 decades. In general, the ensemble mean shows a decrease in the mean annual TC precipitation for the regions mentioned above and an increase by at least 40% for northern Mexico, mainly during the 2020–2030 and 2050–2060 decades (Figure 6). These results demonstrate that the positive changes in summer rainfall over northern Mexico (Figure 3) are essentially due to an increase in TC precipitation for the members rnty and rtty during the 2020–2030 and 2050–2060 decades, as TCs represent up to 60% of the summer rainfall over semiarid Mexican regions [4]. Interestingly, these results agree with the observed positive trends in TC rainfall and TC days over northwestern Mexico found by Dominguez et al. [3] using CHIRPS during the 1981–2017 period.

3.3. Changes in Easterly Wave Features: Seasonal Precipitation, Track Density, and Tropical Cyclogenesis

Mean annual EW track densities were computed as mentioned above for TCs per unit area (1° × 1°). Dominguez et al. [25] examined the EW activity in 10 members of the 24-member ensemble and ERAI for the 1990−2000 period. This study demonstrated that the members rktm, rnty, and rtty simulate EW activity adequately over the MDR when compared to ERAI. However, these members underestimated the maximum EW activity near the West Central American coast (see Figure 10 in Dominguez et al. [25] for further information). This underrepresentation suggests that vortexes that form near this region have such strong winds that are not considered EWs because they quickly evolve to TCs.

Using the previous results provided by Dominguez et al. [25], mean annual anomalies of EW track density were calculated for future climate runs (Figure 7). Over NATL, the members rktm and rnty show significant negative anomalies of EW track density for the southern waveguide for the 2020–2030 and 2030–2040 over the MDR, but, later, the EW activity is projected to scarcely increase by the end of the century (Figure 7). The opposite is shown for the northern waveguide. Intriguingly, the member rtty indicates a higher increment for the southern waveguide during the 2050–2060 and 2080–2090 decades. As a result, the ensemble mean projects a slight increase in EW activity southwards 10° N (Figure 7), which is the waveguide that provides "the seeds" for the tropical cyclogenesis that occurs over MDR [22].

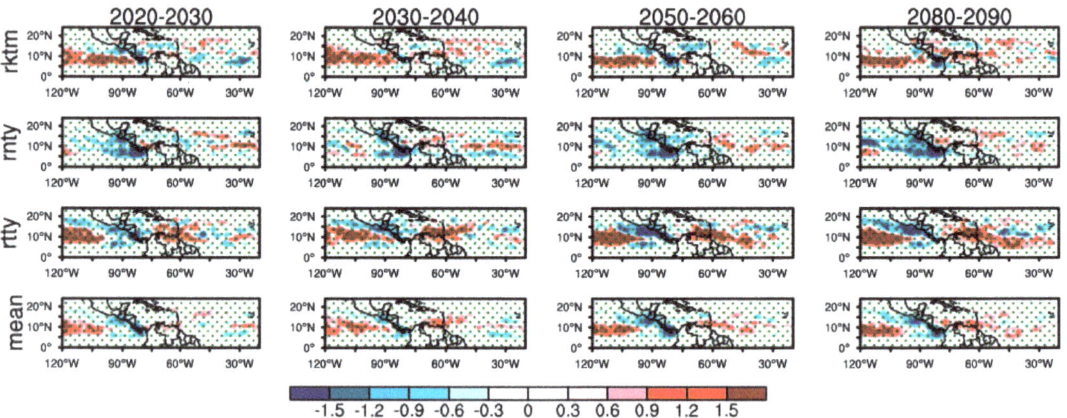

Figure 7. Mean annual anomalies of easterly wave track density in the future ensemble runs with respect to the reference period (1990–2000). The dotted regions indicate a 95% significance level applying the Kolmogorov–Smirnov two-sample test.

Over the EPAC basin, the members rktm and rtty point out a significant shift of the EW activity westwards during all the future decades, but the member rnty shows an opposite signal. In general, the ensemble mean indicates a decrement in EW activity over Central America (where activity is underestimated in the current climate runs) and an increment in the EW activity westwards across the future decades (Figure 7). The latter could have future implications for changes in tropical cyclogenesis locations and the type of TC track that will prevail over this basin, as TC tracks that form far away from continental landmass have a drying effect over land [4].

Dominguez et al. [25] also demonstrated that the members rktm and rnty produced dry EWs, as their associated precipitation was less than 200 mm/yr over EPAC during the 1990–2000 period. However, the rainfall is well represented by the member rtty over the same basin. For the MDR, the three members adequately captured the EW precipitation. Additionally, it is important to note that the three members project a reduction by up to 60% of EW rainfall for the upcoming decades, mainly over the Caribbean region, the Gulf of Mexico, central-southern Mexico, and Central America (not significant). This reduction is the principal cause for the projected decrease in summer rainfall over those regions (Figure 3), as EW contribution to summer rainfall ranges from 50% over the Caribbean region, part of the Gulf of Mexico, and central-southern Mexico, to 60% over Central America [25]. The latter region results are not significant because it seems that the RCM outputs have problems while simulating EW rainfall (Figure 8). This constraint is also shown by the results for TC rainfall over that region (Figure 6).

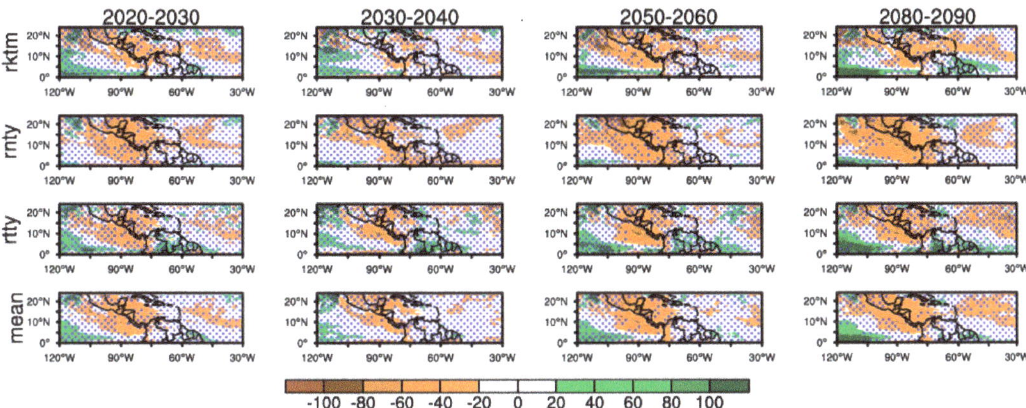

Figure 8. Mean annual percent change (%) in precipitation associated with Easterly Waves in the future ensemble runs with respect to the reference period (1990–2000). The dotted regions indicate a 95% significance level applying the Kolmogorov–Smirnov two-sample test.

In the current climate runs, the members rktm, rnty, and rtty overestimate the mean annual contribution of EWs to summer rainfall, which is obtained from adding all the EW-related precipitation divided by the accumulated precipitation from May to November and then multiplied by 100% to express it in terms of percentage, over the MDR [25]. However, these members tend to underestimate the contribution to summer precipitation only over Central America during the 1990–2000 period (see Figure 13 in Dominguez et al. [25]).

Interestingly, the member rtty shows an increase of 16% in the contribution of EWs to summer rainfall for the 2050–2060 and 2080–2090 decades, mainly over Northern South America (Figure 9). This rise is probably linked to the projected positive change in EW activity during the same decades (Figure 7). On the contrary, the ensemble mean shows a reduction by up to 8% in the contribution to summer rainfall over the Caribbean region and central-southern Mexico for all the upcoming decades (Figure 9). These results support the findings shown in Figures 3 and 8 for the same regions, which suggest that the Caribbean region and Mexico could be affected by drier EWs that could produce less precipitation over the continental landmass. The latter could have implications for regional accumulated rainfall and, consequently, water availability. Besides, the KST shows that the results from rnty and the ensemble mean are not robust for Central America, as future simulations seem to belong to the same distribution of the current climate runs.

Analyzing TCs that come from EWs is important, as 80% of intense hurricanes originate from EWs over NATL [63]. In this study, the percentage of TCs that come from EWs is named tropical cyclogenesis. The hurricane matching technique for the 1990–2000 period shows that 50% of TCs over NATL formed from EWs and 70% of TCs over EPAC originated from these tropical waves [25]. Additionally, Dominguez et al. [25] found that the member rktm simulated 28.8% and 43.9% of tropical cyclogenesis over NATL and EPAC, respectively. The member rnty has similar percentages, 41% and 64.2%, to those observed for the NATL and EPAC, respectively. Notwithstanding, the member rtty poorly captured the tropical cyclogenesis in both basins because it only detected 17.7% and 15.8% of TCs originating from EWs over NATL and EPAC, respectively [25]. In this sense, rktm (rtty) has large biases in capturing tropical cyclogenesis because the difference with the observed percentage is 21.2% (32.3%) for NATL and 26.1% (54.2%) for EPAC.

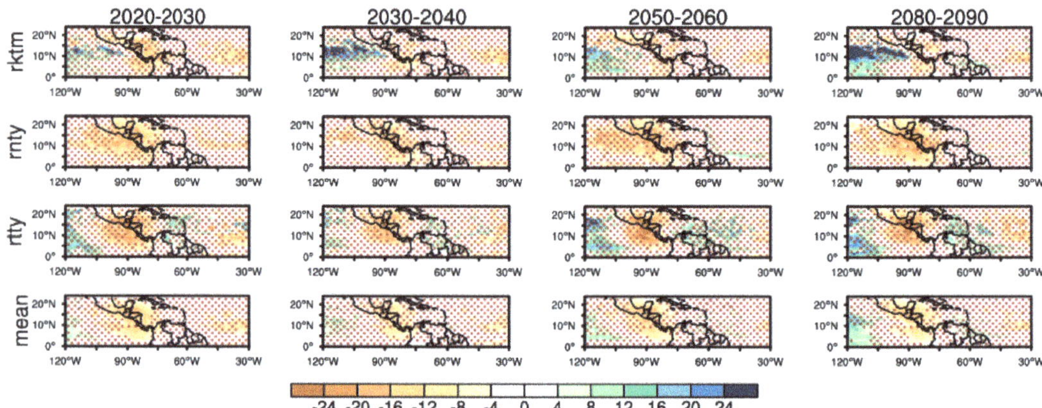

Figure 9. Mean annual percent change (%) in the contribution of Easterly Waves to summer precipitation in the future ensemble runs with respect to the reference period (1990–2000). The dotted regions indicate a 95% significance level applying the Kolmogorov–Smirnov two-sample test.

Even though most of the changes are not significant, the three members and the ensemble mean indicates a decrease by up to 9% in the number of TCs that come from EWs over NATL, except for the member rtty during the 2020–2030 period and for rnty during the 2030–2040 period (Figure 10a), whose simulated tropical cyclogenesis has large biases for the baseline period. These members indicate an increase in the southern waveguide, and further research should be carried out to understand these changes. Surprisingly, the smallest decrement in tropical cyclogenesis is projected for the 2080–2090 decade (Figure 10a). These negative changes for the upcoming decades could probably be associated with unfavorable large-scale environments for tropical cyclogenesis [62]. For the EPAC basin, most of the projected changes are not significant. Tropical cyclogenesis is projected to increase by up to 12% in the members rktm and rtty for the 2020–2030, 2030–240, and 2050–2060 decades, except for the member rnty during the two first decades. However, rktm and rtty have poor skill in simulating tropical cyclogenesis over EPAC. In general, the ensemble mean points out a decrease of 5% in tropical cyclogenesis over this basin during the 2080–2090 period (Figure 10b). Understanding these changes in tropical cyclogenesis using large-scale conditions over NATL and EPAC remains for future work.

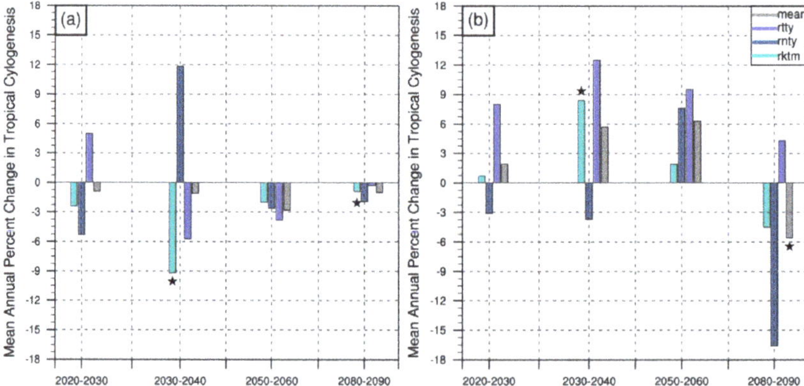

Figure 10. Mean annual percent change in tropical cyclogenesis (%) in the future ensemble runs for: (**a**) the North Atlantic Ocean and (**b**) the Eastern North Pacific Ocean with respect to the reference period (1990–2000). The stars indicate that changes have a 95% significance level applying the Kolmogorov–Smirnov two-sample test.

4. Summary and Conclusions

Tropical Cyclones (TCs) are hazardous tropical phenomena that can produce heavy rainfall and have dangerous winds, causing high economic losses [3]. However, they can also transport substantial amounts of rainfall that induce the recharge of water reservoirs, lakes, and rivers [4]. In that sense, having information about their future regional changes is crucial for decision makers in planning adaptation and mitigation strategies. Easterly Waves (EWs) are also important tropical phenomena, as they are not only precursors of TCs, but also contribute up to 60% of the seasonal rainfall (accumulated rainfall from May to November) over Tropical North America (TNA) [25].

A three-member RCM ensemble was created by NCAR, using multiple physical combinations of parameterizations for the current climate (1990–2000) using the ERAI dataset. The RCM was also driven by the Community Earth System Model (CESM) under the Representative Concentration Pathways 8.5 future emission scenario, which is the most catastrophic in terms of emission concentrations, for creating four future scenarios (2020–2030, 2030–2040, 2050–2060, 2080–2090). Biases from this global model were previously removed following Bruyère et al. [50]. The objective of having three members is to provide enough samples of different atmospheric representations. However, the three-member ensemble does not sample the full range of future possibilities linked to emission scenarios, GCM internal variability, or even different oceanic responses to radiative forcing. Our study is a first step towards looking at the meteorological mechanisms that lead to changes in tropical precipitation due to changes in EWs and TCs. To analyze these changes, we looked at individual ensemble members rather than the ensemble means.

TCs and EWs were tracked using the TRACK technique created by Hodges [51] in the current climate runs and the future climate runs of the three members (rktm, rnty, and rtty). The purpose was to analyze the future changes in EW and TC features: rainfall, track density, contribution to seasonal rainfall, and tropical cyclogenesis.

Our study reveals that the three members project a reduction in the spatial distribution of TC tracks over the Eastern NATL, which is partly consistent with Roberts et al. [15], but these members show slight differences in the spatial distribution over the Gulf of Mexico for all the future decades. Moreover, the three members and the ensemble mean project significant negative changes in TC activity over EPAC during all the future decades (applying the Kolmogorov–Smirnov two-sample test), with the most abrupt reduction in the 105–120° W region and near the Central American coast, where TC activity is overestimated by the members rktm and rtty by the end of the century. In terms of rainfall, while all members project a diminishment of up to 80% of the mean TC precipitation over the Caribbean region, southern Mexico, and Central America (not significant), the ensemble mean indicates an increase of at least 40% in the mean annual TC precipitation over semiarid regions, such as northern Mexico and the southwestern USA, for the four future decades.

Our analysis for future changes in EW characteristics shows that the ensemble mean projects a slight significant increase in EW activity southwards 10° N over NATL, which is the most influential waveguide for the tropical cyclogenesis that occurs over MDR. Over the EPAC basin, the ensemble mean indicates a decrease in EW activity over Central America (not significant), where the current climate runs poorly represent EW activity, and an increment in the EW activity westwards across the future decades. Future changes in EW activity could have implications for shifts in tropical cyclogenesis locations and the type of TC track that will dominate over this basin, as TC tracks that do not make landfall have a drying effect over land [4]. The ensemble mean also projects a reduction by up to 60% of EW rainfall for the upcoming decades, mainly over the Caribbean region, Gulf of Mexico, and central-southern Mexico. What is even more interesting is that these results are also supported by changes in the EW contribution to the seasonal rainfall. The ensemble mean shows a significant reduction by up to 8% in the EW contribution over the Caribbean region and central-southern Mexico for all the upcoming decades. These results suggest that the Caribbean region and Mexico could be affected by drier EWs that could produce less precipitation over land.

The percentage of TCs that originate from EWs over NATL and EPAC (defined here as tropical cyclogenesis) is also projected to change in the upcoming decades. The ensemble mean indicates a decrease by up to 3% in the number of TCs that come from EWs over NATL during the four decades (not significant). For the EPAC basin, tropical cyclogenesis is projected to increase by up to 7% on average for the 2020–2030, 2030–2040, and 2050–2060 decades (not significant). However, the 2080–2090 decade projects a decrease of 5% on average over this basin.

The Kolmogorov–Smirnov two-sample test points out that the RCM outputs have large uncertainty mainly over Central America for the two far future decades (2050–2060 and 2080–2090). Additional RCM simulations are needed to assess if these projected changes in EW and TC features are robust using more future possible scenarios and other RCPs. Furthermore, the changes in the large-scale environments that can conduct to an increase/decrease in tropical cyclogenesis over EPAC/NATL should be further investigated. Another suggestion for future work is to compare our EW rainfall attribution technique with other techniques to provide more robustness in our results.

We conclude that future changes in tropical phenomena characteristics could significantly impact regional precipitation and, thus, future regional water availability. Their variability in the future can determine periods of droughts with high water demand or periods of enough rainfall for agricultural activities. In this sense, regional climate change scenarios should be explored, taking into account changes in the characteristics of tropical phenomena, as they can transport substantial amounts of rainfall. Additionally, future tasks should compare results from the HighResMIP ensemble with RCM outputs to gain an insightful understanding of the information provided in the regional climate change scenarios.

Author Contributions: Conceptualization, C.D.; methodology, C.D., J.M.D., and C.L.B.; validation, C.D.; formal analysis, C.D.; writing—original draft preparation, C.D., J.M.D., and C.L.B.; writing— review and editing, C.D., J.M.D., and C.L.B.; visualization, C.D. All authors have read and agreed to the published version of the manuscript.

Funding: This research received no external funding.

Data Availability Statement: The HURDAT data can be found at https://www.nhc.noaa.gov/data/ (accessed on 13 April 2021). The ERAInterim reanalysis is available at https://www.ecmwf.int/ en/forecasts/datasets/reanalysis-datasets/era-interim (accessed on 13 April 2021). The Reynolds optimum interpolated analysis can be found at https://psl.noaa.gov/data/gridded/data.noaa.oisst. v2.highres.html (accessed on 13 April 2021). The Regional Climate Outputs presented in this study are available on request. The ensemble data are not publicly available due to the fact that it was created under a contract that restricted open data repositories.

Acknowledgments: The authors are grateful to the editor and three anonymous reviewers for their helpful comments and suggestions on the early version of this manuscript. We would like to acknowledge high-performance computing support from Cheyenne provided by NCAR's Computational and Information Systems Laboratory's NCAR Strategic Capability allocation. NCAR is sponsored by the National Science Foundation. C.D. thanks the "Sistema Nacional de Investigadores (SNI)" from CONACyT for her fellowship.

Conflicts of Interest: The authors declare no conflict of interest.

References

1. Hsiang, S.; Camargo, S. Tropical Cyclones: From the Influence of Climate to Their Socioeconomic Impacts. In *Extreme Events: Observations, Modeling, and Economics*; Geophysical Monograph 214; John Wiley & Sons Inc.: Hoboken, NJ, USA, 2016; pp. 303–342. [CrossRef]
2. National Center for Disaster Prevention (CENAPRED). Desastres en México. 2019. Available online: http://www.cenapred.unam. mx/es/Publicaciones/archivos/318-INFOGRAFADESASTRESENMXICO-IMPACTOSOCIALYECONMICO.PDF (accessed on 13 April 2021).

3. Dominguez, C.; Jaramillo, A.; Cuéllar, P. Are the Socioeconomic Impacts Associated with Tropical Cyclones in Mexico Exacerbated by Local Vulnerability and ENSO Conditions? *Int. J. Climatol.* **2021**, *41*, E3307–E3324. [CrossRef]

4. Dominguez, C.; Magaña, V. The Role of Tropical Cyclones in Precipitation Over the Tropical and Subtropical North America. *Front. Earth Sci.* **2018**, *6*, 19. [CrossRef]

5. Zhao, M.; Held, I.M. TC-Permitting GCM Simulations of Hurricane Frequency Response to Sea Surface Temperature Anomalies Projected for the Late-Twenty-First Century. *J. Clim.* **2012**, *25*, 2995–3009. [CrossRef]

6. Murakami, H.; Hsu, P.-C.; Arakawa, O.; Li, T. Influence of Model Biases on Projected Future Changes in Tropical Cyclone Frequency of Occurrence. *J. Clim.* **2014**, *27*, 2159–2181. [CrossRef]

7. Bruyère, C.L.; Rasmussen, R.; Gutmann, E.; Done, J.; Tye, M.; Jaye, A.; Prein, A.; Mooney, P.; Ge, M.; Fredrick, S.; et al. *Impact of Climate Change on Gulf of Mexico Hurricanes*; NCAR Technical Notes; National Center for Atmospheric Research: Boulder, CO, USA, 2017. [CrossRef]

8. Torres-Alavez, J.A.; Glazer, R.; Giorgi, F.; Coppola, E.; Gao, X.; Hodges, K.I.; Das, S.; Ashfaq, M.; Reale, M.; Sines, T. Future Projections in Tropical Cyclone Activity Over Multiple CORDEX Domains from RegCM4 CORDEX-CORE Simulations. *Clim. Dyn.* **2021**. [CrossRef]

9. Bacmeister, J.T.; Wehner, M.F.; Neale, R.B.; Gettelman, A.; Hannay, C.; Lauritzen, P.H.; Caron, J.M.; Truesdale, J.E. Exploratory High-Resolution Climate Simulations Using the Community Atmosphere Model (CAM). *J. Clim.* **2014**, *27*, 3073–3099. [CrossRef]

10. Wehner, M.F.; Reed, K.A.; Li, F.; Prabhat; Bacmeister, J.; Chen, C.-T.; Paciorek, C.; Gleckler, P.J.; Sperber, K.R.; Collins, W.D.; et al. The Effect of Horizontal Resolution on Simulation Quality in the Community Atmospheric Model, CAM5.1. *J. Adv. Model. Earth Syst.* **2014**, *6*, 980–997. [CrossRef]

11. Bacmeister, J.T.; Reed, K.A.; Hannay, C.; Lawrence, P.; Bates, S.; Truesdale, J.E.; Rosenbloom, N.; Levy, M. Projected Changes in Tropical Cyclone Activity under Future Warming Scenarios Using a High-Resolution Climate Model. *Clim. Chang.* **2018**, *146*, 547–560. [CrossRef]

12. Haarsma, R.J.; Roberts, M.J.; Vidale, P.L.; Senior, C.A.; Bellucci, A.; Bao, Q.; Chang, P.; Corti, S.; Fučkar, N.S.; Guemas, V.; et al. High Resolution Model Intercomparison Project (HighResMIP v1.0) for CMIP6. *Geosci. Model. Dev.* **2016**, *9*, 4185–4208. [CrossRef]

13. BAO, Q.; LIU, Y.; WU, G.; HE, B.; LI, J.; WANG, L.; WU, X.; CHEN, K.; WANG, X.; YANG, J.; et al. CAS FGOALS-F3-H and CAS FGOALS-F3-L Outputs for the High-Resolution Model Intercomparison Project Simulation of CMIP6. *Atmos. Ocean. Sci. Lett.* **2020**, *13*, 576–581. [CrossRef]

14. Roberts, M.J.; Camp, J.; Seddon, J.; Vidale, P.L.; Hodges, K.; Vanniere, B.; Mecking, J.; Haarsma, R.; Bellucci, A.; Scoccimarro, E.; et al. Impact of Model Resolution on Tropical Cyclone Simulation Using the HighResMIP–PRIMAVERA Multimodel Ensemble. *J. Clim.* **2020**, *33*, 2557–2583. [CrossRef]

15. Roberts, M.J.; Camp, J.; Seddon, J.; Vidale, P.L.; Hodges, K.; Vannière, B.; Mecking, J.; Haarsma, R.; Bellucci, A.; Scoccimarro, E.; et al. Projected Future Changes in Tropical Cyclones Using the CMIP6 HighResMIP Multimodel Ensemble. *Geophys. Res. Lett.* **2020**, *47*, e2020GL088662. [CrossRef]

16. Knutson, T.R.; Sirutis, J.J.; Vecchi, G.A.; Garner, S.; Zhao, M.; Kim, H.-S.; Bender, M.; Tuleya, R.E.; Held, I.M.; Villarini, G. Dynamical Downscaling Projections of Twenty-First-Century Atlantic Hurricane Activity: CMIP3 and CMIP5 Model-Based Scenarios. *J. Clim.* **2013**, *26*, 6591–6617. [CrossRef]

17. Knutson, T.; Camargo, S.J.; Chan, J.C.L.; Emanuel, K.; Ho, C.-H.; Kossin, J.; Mohapatra, M.; Satoh, M.; Sugi, M.; Walsh, K.; et al. Tropical Cyclones and Climate Change Assessment: Part I: Detection and Attribution. *Bull. Am. Meteorol. Soc.* **2019**, *100*, 1987–2007. [CrossRef]

18. Knutson, T.; Camargo, S.J.; Chan, J.C.L.; Emanuel, K.; Ho, C.-H.; Kossin, J.; Mohapatra, M.; Satoh, M.; Sugi, M.; Walsh, K.; et al. Tropical Cyclones and Climate Change Assessment: Part II: Projected Response to Anthropogenic Warming. *Bull. Am. Meteorol. Soc.* **2020**, *101*, E303–E322. [CrossRef]

19. Serra, Y.L.; Jiang, X.; Tian, B.; Amador-Astua, J.; Maloney, E.D.; Kiladis, G.N. Tropical Intraseasonal Modes of the Atmosphere. *Annu. Rev. Environ. Resour.* **2014**, *39*, 189–215. [CrossRef]

20. Thorncroft, C.; Hodges, K. African Easterly Wave Variability and Its Relationship to Atlantic Tropical Cyclone Activity. *J. Clim.* **2001**, *14*, 1166–1179. [CrossRef]

21. Serra, Y.L.; Kiladis, G.N.; Cronin, M.F. Horizontal and Vertical Structure of Easterly Waves in the Pacific ITCZ. *J. Atmos. Sci.* **2008**, *65*, 1266–1284. [CrossRef]

22. Chen, T.-C.; Wang, S.-Y.; Clark, A.J. North Atlantic Hurricanes Contributed by African Easterly Waves North and South of the African Easterly Jet. *J. Clim.* **2008**, *21*, 6767–6776. [CrossRef]

23. Agudelo, P.A.; Hoyos, C.D.; Curry, J.A.; Webster, P.J. Probabilistic Discrimination between Large-Scale Environments of Intensifying and Decaying African Easterly Waves. *Clim. Dyn.* **2011**, *36*, 1379–1401. [CrossRef]

24. Schreck, C.J.; Molinari, J.; Aiyyer, A. A Global View of Equatorial Waves and Tropical Cyclogenesis. *Mon. Weather Rev.* **2012**, *140*, 774–788. [CrossRef]

25. Dominguez, C.; Done, J.M.; Bruyère, C.L. Easterly Wave Contributions to Seasonal Rainfall over the Tropical Americas in Observations and a Regional Climate Model. *Clim. Dyn.* **2020**, *54*, 191–209. [CrossRef]

26. Crétat, J.; Vizy, E.K.; Cook, K.H. The Relationship between African Easterly Waves and Daily Rainfall over West Africa: Observations and Regional Climate Simulations. *Clim. Dyn.* **2015**, *44*, 385–404. [CrossRef]

27. Janiga, M.A.; Thorncroft, C.D. The Influence of African Easterly Waves on Convection over Tropical Africa and the East Atlantic. *Mon. Weather Rev.* **2016**, *144*, 171–192. [CrossRef]
28. Martin, E.R.; Thorncroft, C. Representation of African Easterly Waves in CMIP5 Models. *J. Clim.* **2015**, *28*, 7702–7715. [CrossRef]
29. Kebe, I.; Diallo, I.; Sylla, M.B.; De Sales, F.; Diedhiou, A. Late 21st Century Projected Changes in the Relationship between Precipitation, African Easterly Jet, and African Easterly Waves. *Atmosphere* **2020**, *11*, 353. [CrossRef]
30. Hsieh, J.-S.; Cook, K.H. A Study of the Energetics of African Easterly Waves Using a Regional Climate Model. *J. Atmos. Sci.* **2007**, *64*, 421–440. [CrossRef]
31. Crosbie, E.; Serra, Y. Intraseasonal Modulation of Synoptic-Scale Disturbances and Tropical Cyclone Genesis in the Eastern North Pacific. *J. Clim.* **2014**, *27*, 5724–5745. [CrossRef]
32. Dee, D.P.; Uppala, S.M.; Simmons, A.J.; Berrisford, P.; Poli, P.; Kobayashi, S.; Andrae, U.; Balmaseda, M.A.; Balsamo, G.; Bauer, P.; et al. The ERA-Interim Reanalysis: Configuration and Performance of the Data Assimilation System. *Q. J. R. Meteorol. Soc.* **2011**, *137*, 553–597. [CrossRef]
33. Landsea, C.W.; Franklin, J.L. Atlantic Hurricane Database Uncertainty and Presentation of a New Database Format. *Mon. Weather Rev.* **2013**, *141*, 3576–3592. [CrossRef]
34. Jiang, H.; Zipser, E.J. Contribution of Tropical Cyclones to the Global Precipitation from Eight Seasons of TRMM Data: Regional, Seasonal, and Interannual Variations. *J. Clim.* **2010**, *23*, 1526–1543. [CrossRef]
35. Khouakhi, A.; Villarini, G.; Vecchi, G.A. Contribution of Tropical Cyclones to Rainfall at the Global Scale. *J. Clim.* **2017**, *30*, 359–372. [CrossRef]
36. Powers, J.G.; Klemp, J.B.; Skamarock, W.C.; Davis, C.A.; Dudhia, J.; Gill, D.O.; Coen, J.L.; Gochis, D.J.; Ahmadov, R.; Peckham, S.E.; et al. The Weather Research and Forecasting Model: Overview, System Efforts, and Future Directions. *Bull. Am. Meteorol. Soc.* **2017**, *98*, 1717–1737. [CrossRef]
37. Reynolds, R.W.; Smith, T.M.; Liu, C.; Chelton, D.B.; Casey, K.S.; Schlax, M.G. Daily High-Resolution-Blended Analyses for Sea Surface Temperature. *J. Clim.* **2007**, *20*, 5473–5496. [CrossRef]
38. Done, J.M.; Holland, G.J.; Bruyère, C.L.; Leung, L.R.; Suzuki-Parker, A. Modeling High-Impact Weather and Climate: Lessons from a Tropical Cyclone Perspective. *Clim. Chang.* **2015**, *129*, 381–395. [CrossRef]
39. Mlawer, E.J.; Taubman, S.J.; Brown, P.D.; Iacono, M.J.; Clough, S.A. Radiative Transfer for Inhomogeneous Atmospheres: RRTM, a Validated Correlated-k Model for the Longwave. *J. Geophys. Res. Atmos.* **1997**, *102*, 16663–16682. [CrossRef]
40. Kain, J.S.; Fritsch, J.M. A One-Dimensional Entraining/Detraining Plume Model and Its Application in Convective Parameterization. *J. Atmos. Sci.* **1990**, *47*, 2784–2802. [CrossRef]
41. Han, J.; Pan, H.-L. Revision of Convection and Vertical Diffusion Schemes in the NCEP Global Forecast System. *Weather Forecast.* **2011**, *26*, 520–533. [CrossRef]
42. Tiedtke, M. A Comprehensive Mass Flux Scheme for Cumulus Parameterization in Large-Scale Models. *Mon. Weather Rev.* **1989**, *117*, 1779–1800. [CrossRef]
43. Thompson, G.; Field, P.R.; Rasmussen, R.M.; Hall, W.D. Explicit Forecasts of Winter Precipitation Using an Improved Bulk Microphysics Scheme. Part II: Implementation of a New Snow Parameterization. *Mon. Weather Rev.* **2008**, *136*, 5095–5115. [CrossRef]
44. Janjić, Z.I. The Step-Mountain Eta Coordinate Model: Further Developments of the Convection, Viscous Sublayer, and Turbulence Closure Schemes. *Mon. Weather Rev.* **1994**, *122*, 927–945. [CrossRef]
45. Hong, S.-Y.; Noh, Y.; Dudhia, J. A New Vertical Diffusion Package with an Explicit Treatment of Entrainment Processes. *Mon. Weather Rev.* **2006**, *134*, 2318–2341. [CrossRef]
46. Chen, F.; Dudhia, J. Coupling an Advanced Land Surface–Hydrology Model with the Penn State–NCAR MM5 Modeling System. Part I: Model Implementation and Sensitivity. *Mon. Weather Rev.* **2001**, *129*, 569–585. [CrossRef]
47. Hurrell, J.W.; Holland, M.M.; Gent, P.R.; Ghan, S.; Kay, J.E.; Kushner, P.J.; Lamarque, J.-F.; Large, W.G.; Lawrence, D.; Lindsay, K.; et al. The Community Earth System Model: A Framework for Collaborative Research. *Bull. Am. Meteorol. Soc.* **2013**, *94*, 1339–1360. [CrossRef]
48. Taylor, K.E.; Stouffer, R.J.; Meehl, G.A. An Overview of CMIP5 and the Experiment Design. *Bull. Am. Meteorol. Soc.* **2012**, *93*, 485–498. [CrossRef]
49. Moss, R.H.; Nakicenovic, N.; O'Neill, B.C. Towards New Scenarios for Analysis of Emissions, Climate Change, Impacts, and Response Strategies. In Proceedings of the IPCC Expert Meeting on New Scenarios, Noordwijkerhout, The Netherlands, 19–21 September 2007; Moss, R., Intergovernmental Panel on Climate Change, Eds.; IPCC Expert Meeting Report; Intergovernmental Panel on Climate Change: Geneva, Switzerland, 2008; ISBN 978-92-9169-125-8.
50. Bruyère, C.L.; Done, J.M.; Holland, G.J.; Fredrick, S. Bias Corrections of Global Models for Regional Climate Simulations of High-Impact Weather. *Clim. Dyn.* **2014**, *43*, 1847–1856. [CrossRef]
51. Hodges, K.I. Feature Tracking on the Unit Sphere. *Mon. Weather Rev.* **1995**, *123*, 3458–3465. [CrossRef]
52. Walsh, K.J.E.; Fiorino, M.; Landsea, C.W.; McInnes, K.L. Objectively Determined Resolution-Dependent Threshold Criteria for the Detection of Tropical Cyclones in Climate Models and Reanalyses. *J. Clim.* **2007**, *20*, 2307–2314. [CrossRef]
53. Serra, Y.L.; Kiladis, G.N.; Hodges, K.I. Tracking and Mean Structure of Easterly Waves over the Intra-Americas Sea. *J. Clim.* **2010**, *23*, 4823–4840. [CrossRef]

54. Wilks, D. Statistical Methods in the Atmospheric Sciences. In *International Geophysics*, 3rd ed.; Elsevier: Amsterdam, The Netherlands; Academic Press: Boston, MA, USA, 2011; Volume 100, ISBN 978-0-12-385022-5.
55. Zhang, Y.; Fueglistaler, S. Mechanism for Increasing Tropical Rainfall Unevenness with Global Warming. *Geophys. Res. Lett.* **2019**, *46*, 14836–14843. [CrossRef]
56. Murakami, H.; Vecchi, G.A.; Underwood, S.; Delworth, T.L.; Wittenberg, A.T.; Anderson, W.G.; Chen, J.-H.; Gudgel, R.G.; Harris, L.M.; Lin, S.-J.; et al. Simulation and Prediction of Category 4 and 5 Hurricanes in the High-Resolution GFDL HiFLOR Coupled Climate Model. *J. Clim.* **2015**, *28*, 9058–9079. [CrossRef]
57. Emanuel, K. Response of Global Tropical Cyclone Activity to Increasing CO2: Results from Downscaling CMIP6 Models. *J. Clim.* **2021**, *34*, 57–70. [CrossRef]
58. Gutmann, E.D.; Rasmussen, R.M.; Liu, C.; Ikeda, K.; Bruyere, C.L.; Done, J.M.; Garrè, L.; Friis-Hansen, P.; Veldore, V. Changes in Hurricanes from a 13-Yr Convection-Permitting Pseudo-Global Warming Simulation. *J. Clim.* **2018**, *31*, 3643–3657. [CrossRef]
59. Saha, S.; Moorthi, S.; Pan, H.-L.; Wu, X.; Wang, J.; Nadiga, S.; Tripp, P.; Kistler, R.; Woollen, J.; Behringer, D.; et al. The NCEP Climate Forecast System Reanalysis. *Bull. Am. Meteorol. Soc.* **2010**, *91*, 1015–1058. [CrossRef]
60. Durán-Quesada, A.M.; Sorí, R.; Ordoñez, P.; Gimeno, L. Climate Perspectives in the Intra-Americas Seas. *Atmosphere* **2020**, *11*, 959. [CrossRef]
61. Imbach, P.; Locatelli, B.; Zamora, J.C.; Fung, E.; Calderer, L.; Molina, L.; Ciais, P. Impacts of Climate Change on Ecosystem Hydrological Services of Central America: Water Availability. In *Climate Change Impacts on Tropical Forests in Central America: An Ecosystem Service Perspective*; Aline, C., Ed.; Routledge Publishing: New York, NY, USA, 2015; ISBN 978-0-415-72080-9.
62. Jaye, A.B.; Bruyère, C.L.; Done, J.M. Understanding Future Changes in Tropical Cyclogenesis Using Self-Organizing Maps. *Weather Clim. Extrem.* **2019**, *26*, 100235. [CrossRef]
63. Landsea, C.W. A Climatology of Intense (or Major) Atlantic Hurricanes. *Mon. Weather Rev.* **1993**, *121*, 1703–1713. [CrossRef]

Article

Demonstration of the Temporal Evolution of Tropical Cyclone "Phailin" Using Gray-Zone Simulations and Decadal Variability of Cyclones over the Bay of Bengal in a Warming Climate

Prabodha Kumar Pradhan [1,*], Vinay Kumar [2], Sunilkumar Khadgarai [3], S. Vijaya Bhaskara Rao [1], Tushar Sinha [2], Vijaya Kumari Kattamanchi [1] and Sandeep Pattnaik [4]

[1] Department of Physics, Sri Venkateswara University, Tirupati 517 502, India; drsvbr.acas@gmail.com (S.V.B.R.); vijayakattamanchi@gmail.com (V.K.K.)

[2] Department of Environmental Engineering, Texas A & M University, Kingsville, TX 78363, USA; vinay.kumar@tamuk.edu (V.K.); Tushar.Sinha@tamuk.edu (T.S.)

[3] Indian Institute of Tropical Meteorology, Ministry of Earth Sciences, Pune 411 001, India; sunil.khadgarai@tropmet.res.in

[4] Earth, Ocean and Climate Sciences, Indian Institute of Technology, Bhubaneswar 751 013, India; spt@iitbbs.ac.in

* Correspondence: prabodha.svu@gmail.com

Citation: Pradhan, P.K.; Kumar, V.; Khadgarai, S.; Rao, S.V.B.; Sinha, T.; Kattamanchi, V.K.; Pattnaik, S. Demonstration of the Temporal Evolution of Tropical Cyclone "Phailin" Using Gray-Zone Simulations and Decadal Variability of Cyclones over the Bay of Bengal in a Warming Climate. *Oceans* **2021**, *2*, 648–674. https://doi.org/10.3390/oceans2030037

Academic Editor: Hiroyuki Murakami

Received: 6 April 2021
Accepted: 31 August 2021
Published: 10 September 2021

Publisher's Note: MDPI stays neutral with regard to jurisdictional claims in published maps and institutional affiliations.

Abstract: The intensity and frequency variability of cyclones in the North Indian Ocean (NIO) have been amplified over the last few decades. The number of very severe cyclonic storms (VSCSs) over the North Indian Ocean has increased over recent decades. "Phailin", an extreme severe cyclonic storm (ESCS), occurred during 8–13 October 2013 over the Bay of Bengal and made landfall near the Gopalpur coast of Odisha at 12 UTC on 12 October. It caused severe damage here, as well as in the coastal Odisha, Andhra Pradesh, and adjoining regions due to strong wind gusts (~115 knot/h), heavy precipitation, and devastating storm surges. The fidelity of the WRF model in simulating the track and intensity of tropical cyclones depends on different cloud microphysical parameterization schemes. Thus, four sensitivity simulations were conducted for Phailin using double-moment and single-moment microphysical (MP) parameterization schemes. The experiments were conducted to quantify and characterize the performance of such MP schemes for Phailin. The simulations were performed by the advanced weather research and forecasting (WRF-ARW) model. The model has two interactive domains covering the entire Bay of Bengal and adjoining coastal Odisha on 25 km and 8.333 km resolutions. Milbrandt–Yau (MY) double-moment and WRF single-moment microphysical schemes, with 6, 5, and 3 classes of hydrometeors, i.e., WSM6, WSM5, and WSM3, were used for the simulation. Experiments for Phailin were conducted for 126 h, starting from 00 UTC 8 October to 06 UTC 13 October 2013. It was found that the track, intensity, and structure of Phailin are highly sensitive to the different microphysical parameterization schemes. Further, the precipitation and cloud distribution were studied during the ESCS stage of Phailin. The microphysics schemes (MY, WSM3, WSM5, WSM6), along with Grell–Devenyi ensemble convection scheme predicted landfall of Phailin over the Odisha coast with significant track errors. Supply of moisture remains a more crucial component than SST and wind shear for rapid intensification of the Phailin 12 h before landfall over the Bay of Bengal. Finally, the comparison of cyclone formation between two decades 2001–2010 and 2011–2020 over the Bay of Bengal inferred that the increased numbers of VSCS are attributed to the supply of abundant moisture at low levels in the recent decade 2011–2020.

Keywords: very severe cyclonic storm; Phailin; microphysics; hydrometeors; WRF model; Bay of Bengal

1. Introduction

During the post-monsoon season, tropical cyclones are the primary source for clouds and precipitation over the tropics [1–5]. It has been reported that clouds are the essential meteorological element in the formation and structure of tropical cyclones (TCs). The well-organized clusters of convective clouds around the central area of surface low-pressure over the sea surface help to develop the TC [6–9]. The energy required for a TC to intensify is acquired from the direct transfer of sensible and latent heat fluxes from the warm ocean surface via the convection process. The cloud bands in the inner regions of TC are mostly cumulonimbus in nature and happen to be located within the spinning vortex and intricately connected with the dynamics of the cyclone itself. Houze (2010) reported that as high-resolution core physics models become more widely used, forecasting the probabilities of extreme weather and heavy precipitation at specific times and locations will become a feasible goal for land-falling cyclones. Understanding the microphysical processes associated with TCs, particularly in the tropics, is challenging and remains unclear [10]. The microphysical parameterization (MP) is an important source of uncertainty in numerical weather predictions of mesoscale convective systems [11]. Due to the complexity of microphysical processes, various MP schemes have been developed over the past decades based on Eulerian approaches to represent cloud and precipitation in mesoscale models. For simulation of TCs over the Bay of Bengal (BoB), cloud microphysics schemes, such as Kessler, WSM3/5/6, Ferrier, Godard, Thompson, Milbrandt–Yau, Morrison, WDM, and Lin, have been widely used in the WRF model. However, single-moment bulk microphysical schemes calculate the mixing ratio categories (cloud, water, rain, cloud ice, snow, and graupel). In contrast, the double-moment bulk microphysical scheme predicts the corresponding mixing ratio number with concentration and mass mixing ratios [12]. Further, there are differences in the number of anticipated moments, while such MP schemes follow gamma distribution for precipitating hydrometers. Hong et al., (2004) concluded that the WSM3 simple ice scheme without mixed phase can capture the warm rain processes better than mixed-phase schemes such as WSM5 and WSM6 over tropical regions [13]. They further concluded that the simple ice scheme without mixed-phase microphysics is good enough to resolve the mesoscale features in 25 km grid resolution. The horizontal grid spacing between around 5 and 12 km is referred to as the convective gray-zone resolution (hereafter referred to as gray-zone resolution for short), which avoids convection scheme uncertainties as the results rely on the cloud microphysics and PBL schemes [14].

Clouds associated with TCs are typically organized into large rings and bands. TCs have clouds and precipitation structures that are similar to the mesoscale convective systems over mid-latitude [7]. Therefore, similarly to improving numerical weather prediction (NWP) models for quantitative precipitation forecasts, the problem of improving TC forecasts from NWP is closely related to how to better simulate microphysics of winds, rainfall, and moisture. It has been recognized that experiments with various complex microphysical processes could significantly influence the TC's intensity, structure, and evolution at finer horizontal and vertical resolutions [9,13]. The TCs are among the most devastating extreme weather phenomena in the tropics. Models' skills in predicting the TC track and intensity over the Bay of Bengal (BoB) have been discussed in detail by previous studies [6,15,16]. There are numerous MP schemes available in weather research and forecasting (WRF) models [17]. Most of these schemes, such as LIN explicit, WSM6, Thompson, WDM6, Thompson and Morrison, WSM 3-class simple ice, and Ferrier have been widely used for TC forecasts over the North Indian Ocean and the results are discussed in [18–22]. The parameterized production rate of hydrometeors is the crucial aspect for the sensitivity of the quantitative precipitation forecast (QPF) to different MP schemes [9,23,24]. It is important to provide an accurate QPF during the severe cyclogenesis stage and at the land-fall stage of Phailin or any other cyclone. The simulation of different stages of a TC is very much associated with the convection and cloud process. Thus, using various MP schemes, the models can accurately capture the natural variability and dynamics of cloud processes [11]. During the cyclone evolution, the intensity and track forecast largely depend on the diabatic heating rates [25,26], which arise mainly by latent heat release by the condensation processes within the system. However, the diabatic heating rates induce mixing over the convective and stratiform rainfall regions. The vertical profiles of diabatic heating rates and microphysical properties of clouds further define the diurnal and temporal extent of the cyclone [25].

In post-monsoon season, especially in October, the number of severe to super severe cyclone formations remains the highest and quite strongly affects the Indian subcontinent, Bangladesh, Myanmar, Sri Lanka, Oman, Somalia, and Yemen. In the latest decade (2011–2020) the number of severe cyclones over the Indian Ocean has increased, while the number of depressions has decreased. A few of the intensified and severely damaging cyclones (of a total nearly in the billions) were Gonu, Nargis, Giri, Thane, Phailin, Nilofar, Vardah, Ockhi, Mekunu, Fani, Kyarr, and the latest super cyclone Amphan in 2020 (IMD Archives). Over the last few decades, two ESCSs (Odisha, 1999; Phailin, 2013) have made landfall over the Odisha coast. The VSCS occurred during the post-monsoon season and caused socio-economic damages and casualties [27,28].

The simulation studies of the Phailin cyclone were done by [19,29–31], but none of the studies paid attention to the cloud microphysical aspects of the Phailin cyclone. However, the microphysical processes and characteristics of different hydrometeor concentrations may be considered as a decisive factor for TC intensity prediction [18]. In the present study, the sensitivity experiments of various MP parameterization schemes with 25 km and 8.333 km model resolutions were used to highlight the track and intensity prediction of Phailin. The objective of this study was to investigate the sensitivity of single and double-moment MP schemes in the WRF model to simulate the track, intensity, circulation dynamics in the eyewall, precipitation, and vertical cross-section of extreme severe cyclonic storm Phailin (ESCS). Thus, the present study discusses the track, intensity, and dynamical mechanisms that are responsible for the eyewall formation, rainfall, and characteristics of hydrometeors. In this article, synoptic conditions of Phailin are discussed in Section 2, details of numerical experiments and data used are discussed in Section 3, results and discussions are presented in Section 4. Recent changes in TCs formation over the BoB are presented in Section 5, and, finally, the essential findings based on the results are discussed in Section 6.

2. The Synoptic Conditions during Phailin Cyclone

The Phailin cyclone with a lifespan of 6 days (8–13 October 2013) initially originated as a depression near Andaman and Nicobar Island (in BoB, India). The pressure (hPa) and wind speed (knots) corresponding to different stages of Phailin are represented in Table 1. A low-pressure system that occurred on 8 October 2013 over the BoB was named Phailin. The Phailin initially started as depression and reached up to a very super cyclonic storm (VSCS) on 06 UTC of 10 October 2013. Initially, the low-pressure system lingered over the same location 12° N–96° E as a depression (D) for a day. The system got intensified into a deep depression (DD) in the next 24 h and then moved in the northwest direction.

Table 1. Synoptic conditions of Phailin during 8–13 October 2013 obtained from IMD report.

Stage	Time	Pressure (hPa)	Wind Speed (kts)	Duration (Hours)/ Shape of the System	Latitude/Longitude
Depression (D)	03 UTC 8 Oct 2013	1004	25	24 Dense clouds	12.0° N; 96.0° E
Deep Depression (DD)	03 UTC 9 Oct 2013	1001	30	09 Dense clouds	13.0° N; 93.5° E
Cyclonic Storms (CS)	12 UTC 9 Oct 2013	999	35	12 Dense clouds	13.5° N;92.5° E
Severe cyclonic storm (SCS)	03 UTC 10 Oct 2013	990	55	03 Very dense clouds	14.5° N; 91.0° E
Very severe cyclonic storm (VSCS)	06 UTC 10 Oct 2013	984	65	21 Almost closed eye and fair	15.9° N; 90. 5° E

Table 1. *Cont.*

Stage	Time	Pressure (hPa)	Wind Speed (kts)	Duration (Hours)/ Shape of the System	Latitude/Longitude
Extreme severe cyclonic storm (ESCS)	03 UTC 11 Oct 2013	940	115	20 Almost closed eye and good	16.0° N; 88.5° E
Extreme severe cyclonic storm (ESCS)	03 UTC 12 Oct 2013	940	115	24 Almost closed eye and more prominent	17.8° N; 86.0° E
Severe cyclonic storm (SCS)	03 UTC 13 Oct 2013	990	55	Almost closed Eye	17.8° N; 85.9° E

This DD further intensified within the next 12 h as a cyclonic storm (CS) at 12 UTC, 9 October 2013 (IMD, 2013). Phailin became a VSCS at 06 UTC on 10 October 2013, located at 14.5° N, 91.0° E, and stirred northwestwards with a maximum sustained wind speed of 55 knots and 990 hPa. Due to the vertical wind shear of 5–10 knots, the cyclonic storm became a VSCS with a wind magnitude of 70 knots, and pressure dropped 26 hPa off 990 hPa on 10 October 2013. The VSCS persisted over the middle of the BOB for 24 h, with a central sea level pressure (CSLP) of 940 hPa and wind speed of 115 knots, and continued to move in the northwest direction until it gradually became an ESCS. Figure 1 illustrates the double eyewall of ESCS Phailin at 00 UTC, 12 October 2013. To evaluate the tropical cyclone's structural characteristics, the satellite images derived from the water vapor and rain rate of Phailin are shown in Figure 1a,b. The rain rates were derived using a combination of passive microwave channels (F-17) representing the core rain bands and eyewall decoration of Phailin during the ESCS stage. The eyewall represented the consolidated system and underwent a second eyewall formation, as shown in Figure 1b with a categorization of 5 in terms of intensity. The system subsequently made landfall later that day (13 October) near the Gopalpur coast of Odisha around 22:30 IST (17 UTC), near peak intensity, which caused severe damages as discussed [18]. The Phailin with VSCS intensity passed through the Gopalpur coast of Odisha and adjoining Andhra Pradesh at 15 UTC on 12 October 2013 at latitude 19.2° N and 84.9° E (IMD, 2013). After 24 h of landfall, the intensity of the ESCS declined and turned into an SCS, holding a CSLP of 990 hPa and wind speed of 55 kt. Satellite images of Phailin on 24 h temporal intervals were used to determine whether the TC intensified, weakened, or retained intensity, as followed by the Dvorak method (1975). Therefore, the NOAA satellite pictures were considered at two different stages of Phailin at 24 h intervals, as represented in Figure 1a,b. The rain rate patterns identified for the eyewall development for Phailin were valid on 11 October 2013 at 23:30 UTC (Figure 1c), which rapidly intensified into a further severe cyclone. Figure 1d shows an intensive rain rate with a well-organized spiral eyewall that was maintained by cold clouds at the upper troposphere surrounded by warm temperatures in the eye region.

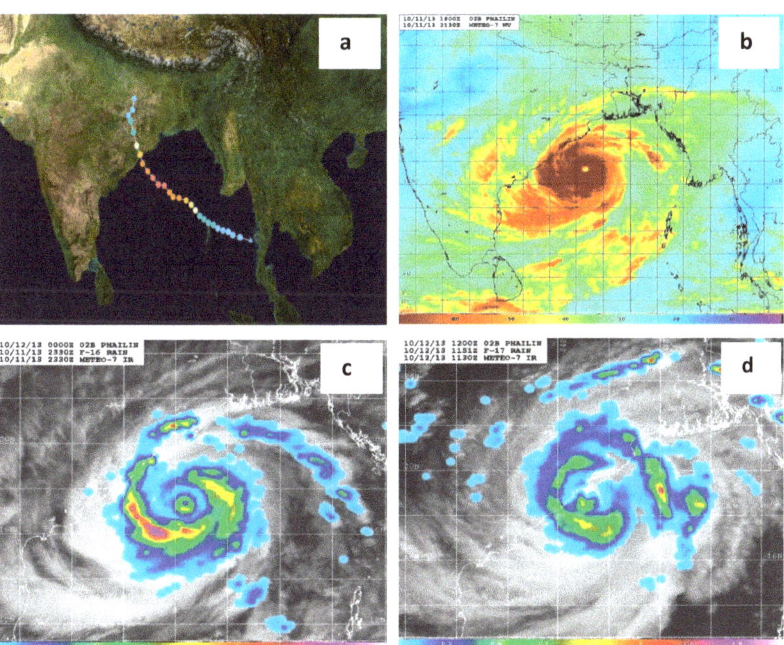

Figure 1. (**a**) Track of Phailin during 8–13 October 2013; according to SSHWS, the system became equivalent to a Category 5 (NASA) on 00 UTC 12 October 2013; Satellite images METEO 7 of ESCS Phailin, (**b**) water vapor 00 UTC 12 October 2013, (**c**) rain rate valid on 11 October 2013 at 23:30 UTC, (**d**) rain rate after landfall valid on 13 October 2013 at 11:30 UTC shown double eyewall features [for details https://rammb-data.cira.colostate.edu/tc_realtime/, (accessed on 21 August 2021)].

3. Numerical Experiments and Data Used

3.1. WRF Model Setup

The model used in this study was the advanced research WRF (ARW), version 3.7.1 [17]. The analyses and 6 h forecast fields of the final analysis (FNL) of the NCEP at $1.0° \times 1.0°$ grid space were taken as the initial and boundary conditions for the ARW model. The lateral boundary conditions were updated in a 6 h intervals, and the SST was kept constant throughout the model integration. The United States Geological Survey (USGS) data with 10 min and 5 min resolution were used to provide permanent land surface fields, such as terrain/topography. A double domain of 25 km and 8.333 km (gray zone simulations) were chosen, which extended from 75–110° E and 4–32° N with 42 vertical levels [14]. The vertical levels were closely placed in the lower levels (12 levels below 850 hPa and 22 levels between 850–500 hPa) and were relatively coarser in higher levels. The domains are presented in Figure 2. The model was integrated at a 3 h interval using the Yonsei University [32] (YSU) planetary boundary layer (PBL) scheme, with Grell and Devenyi ensemble [33] (GDE) for the convective parameterization scheme. For TC simulation, the GDE scheme has the least errors in terms of the intensity of tropical cyclones, as discussed in [34,35]. Thus, GDE convective scheme was used in this study. The thermal diffusion (slab) scheme [36] was used for the land surface representation in the WRF model.

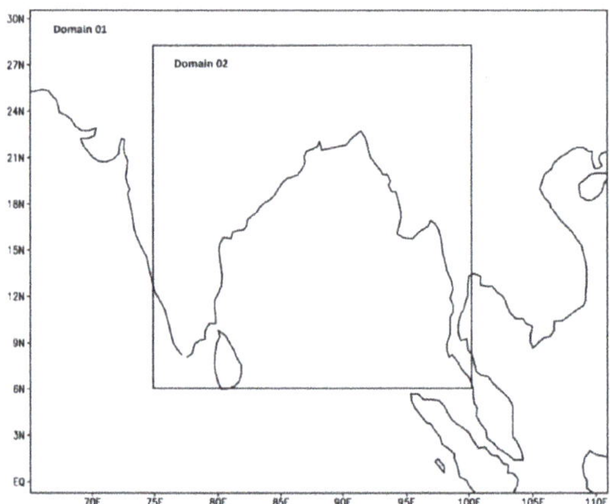

Figure 2. WRF model domains used for experiments of Phailin and horizontal resolutions of D01 (25 km) and D02 (8.333 km, gray-zone grid), respectively.

The rapid radiative transfer model (RRTM) longwave radiation scheme and the shortwave radiation scheme of [37] were used to simulate the radiative forcing. To explore the sensitivity of cloud microphysical parameterization schemes, four MP schemes were chosen to perform the experiments up to 126 h, and the model's outputs were generated after 3 h intervals. The details of domain and physics options used in the model experiments are represented in Table 2. The MP schemes and their characteristics are discussed in Table 3.

3.2. Classification of Single- and Double-Moment Microphysics

The primary microphysical species are water vapor, cloud droplets, rain droplets, cloud ice crystals, snow, rimed ice, graupel, and hail. Microphysics budgets depend on atmospheric dynamical and thermodynamical conditions, which determine the partitioning of hydrometeors [38]. Most of the schemes may have two or three ice categories; however, the degree of sophistication used to represent the microphysics processes varies considerably [39]. There has been rapid progress in the understanding of cloud microphysical processes in recent decades, and many microphysical schemes have been developed for applications in NWP and climate models. Thus, cloud processes can be studied with more confidence, especially from the point of view of linking cloud-scale processes to large-scale atmospheric circulations [40]. Numerous studies have discussed a couple of atmospheric models that applied an Eulerian approach for the cloud and thermodynamic variables, not only for the temperature and water vapor but also for the prognostic variables, such as ice particles, which occur as sparsely distributed liquid drops and ice particles [41,42]. The detailed descriptions and formulation of hydrometeors are illustrated in Table 3.

Table 2. WRF model and domain configurations.

Model Features	Non-Hydrostatic
Version	3.7.1
Horizontal resolution	25 km, 8.333 km
Vertical levels	42
Topography	USGS
Dynamics	
Time integration	3rd order Runga-Kutta
Time steps	30 s
Horizontal grid distribution	Arakawa C-grid
Spatial differencing scheme	6th order centered differencing
Physics	
Radiation scheme	Dudhia for short wave radiation/RRTM longwave radiation
Surface layer	Monin–Obukhov similarity theory
Land surface parameterization	5-layer thermal diffusion
PBL parameterization scheme	Yonsei University scheme (Hong et al. 2006)
Cumulus parameterization scheme	Grell-Devenyi Ensemble (GDE)
Cloud microphysics	(1) Milbrandt–Yau double-moment 7-class (MY) (2) WSM6-class (3) WSM5-class (4) WSM3-class
Initial and boundary conditions	Real data from NCEP FNL (1 × 1 degree)

WSM3: The WRF model is a community model suitable for research and forecasting [43,44]. In this scheme, the water mixing ratios are prognostic and single-moment in nature. In the WRF model, the modifications of the microphysics of clouds and precipitation are implemented as NCEP simple ice (three classes: vapor, cloud/ice, and rain/snow), referred to as WSM3 scheme, and details are discussed [45]. *WSM5*: The mixed-phase (five classes: vapor, cloud, ice, rain, and snow) schemes are referred to as WSM5 schemes. Hong et al. (2004) suggested that there is a significant role of the microphysical properties on mesoscale forecast [46]. They further added that the simple ice scheme without mixed-phase microphysics is enough to resolve mesoscale features on a 25 km grid resolution. The modifications in the ice microphysical processes result in a realistic distribution of clouds through auto-conversion of cloud water to rain, similar to Kessler's formula [47]. *WSM6*: The WSM6 scheme has been developed by adding additional processes related to graupel to the WSM5 scheme [48]. *Milbrandt and Yau (MY)*: A bulk parameterization microphysics scheme in atmospheric models is important to develop details on rainfall and other features of a system. MY [49,50] is a computationally efficient scheme, thus, it is widely used to understand the strength and limitations of various rain-bearing processes.

Table 3. WRF v3.7.1. Cloud microphysical schemes used for the Phailin experiment.

mp_Physics	Microphysical Scheme Name	Abbreviation	Hydrometeors
9	Milbrandt–Yau double-moment 7-class [Milbrandt and Yau 2005; Milbrandt and Yau 2005]	MY	vapor, cloud, rain, ice, snow, graupel, and hail
6	WRF single-moment 6-class [Hong et al., 2006]	WSM6	vapor, cloud, rain, ice, snow, and graupel
4	WRF single-moment 5-class [Hong et al., 2004]	WSM5	vapor, cloud, rain, ice, and snow
3	WRF single-moment 3-class [Hong et al., 2004]	WSM3	vapor, cloud/ice, and rain/snow.

4. Results and Discussion

4.1. Intensity and Structure of Phailin

Evaluation of track and intensity is given priority for the precise forecasting of low-pressure systems. The simulated track positions of Phailin from different MP experiments and IMD observations are represented in Figure 3. It was noticed that the MP experiments simulated slight deviation in tracks during the genesis stage of the Phailin.

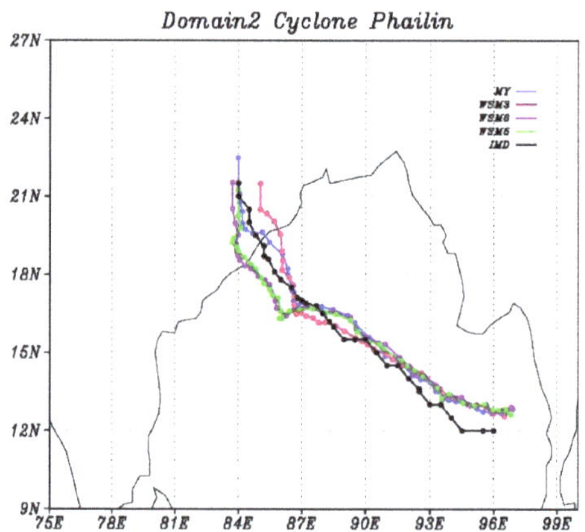

Figure 3. Model-simulated track positions of Phailin during 00 UTC, 8 October to 06 UTC, 13 October 2013 using 00 UTC, 8 October 2013 as initial conditions. The positions were identified through the minimum sea level pressure from different WRF experiments along with IMD observations.

The simulated track positions from DD to VSCS have deviated to the east and north of the IMD observation. However, following later landfall of VSCS, positions of simulated tracks were located west of the IMD observed track. From the model experiments, the initial location of the simulated tracks may have shifted relative to observations [39,51]. The track errors from different MP scheme experiments were drawn using the formula from [52] and are shown in Figure 4. The simulated track errors from the MP experiments from domain 1 (D01: 25 km) and domain 2 (D02: 8.333 km) resolutions indicate the sensitivity of tracks in terms of resolutions. The track errors during the simulations were found to vary, as shown in Figure 4a,b. The average number track errors over 12 h was lower in D02 as compared to D01, especially for the MY scheme, which was expected from MP experiments and may be due to the changes in explicit moisture processes. Overall, among the four MP experiments, the WSM3 scheme showed less track errors than MY, WSM6, and WSM5 schemes during the simulations in both domains. During the simulation period, WSM6 and WSM5 showed more track errors compared to MY and WSM3 schemes because the moisture process was being resolved accurately in the latter schemes.

WRF-simulated MSLP and 10 m wind field of Phailin based on 00 UTC, 08 October 2013 initial conditions along with IMD observations are represented in Figure 5a,b. To understand the different stages of Phailin, some of the important parameters were analyzed such as center sea level pressure (CSLP) and maximum sustained wind (MSW) evolution during 8–13 October 2013. For Phailin, the observed estimates indicate that the lowest pressure drops and MSW were about 940 hPa and 115 knots, respectively, at 00 UTC, 13 October 2013. Though the simulation experiments using MY, WSM6, WSM5, and WSM3 schemes underestimated the peak pressure drops, the evolutions expressively agreed with IMD values.

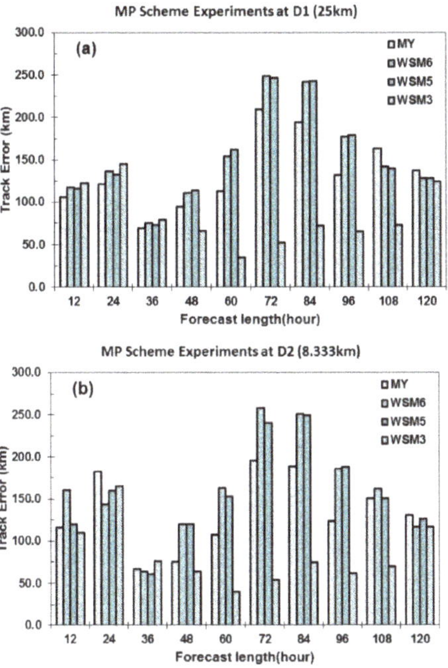

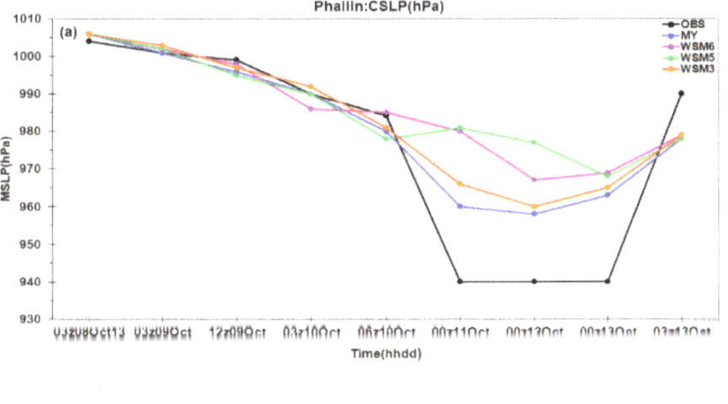

Figure 4. Simulated track errors from the 12 h average derived from different WRF experiments at (**a**) D01 and (**b**) D02 resolutions. The bar and scheme are in the sequence of MY, WSM6, WSM5, and WSM3 respectively.

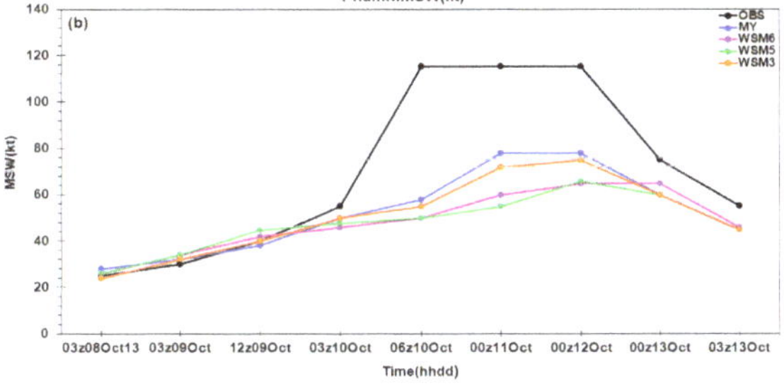

Figure 5. Intensity simulation of (**a**) MSLP (hPa) and (**b**) 10 m maximum sustained wind (MSW; knots) derived from IMD and WRF model experiments during different stages of Phailin at 8.333 km resolution.

As per the IMD report, during a 72 h period, the low-pressure system intensified and reached the stage of VSCS with a wind speed of 60–100 knots (IMD, 2013) at 00 UTC, 11 October, whereas simulations with MP schemes could not predict similar intensity. The MY scheme could simulate maximum sustained wind of 55 knots with the lowest CSLP of 970 hPa, as depicted in Figure 5a. For wind field, MY and WSM3 MP schemes captured maximum wind of 55 knots, whereas WSM6 and WSM5 MP schemes were unable to simulate the same feature. Furthermore, WSM3, WSM5, and WSM6 MP schemes carry biases in simulating the wind speed, as simulated by the MY scheme. The RMSE of CSLP (hPa) and MSW (knots) double-moment MY schemes showed relatively fewer errors than WSM6 and WSM5 in D02 experiments, as illustrated in Table 4.

Table 4. The 24-h average RMSE of CSLP (hPa) and MSW (knots) for the Phailin cyclone derived from D02 experiments.

Stage of Phailin	Simulation Length	RMSE of CSLP (hPa)				RMSE of Wind at 10-m (Knots)			
		MY	WSM6	WSM5	WSM3	MY	WSM6	WSM5	WSM3
D	03z08	1.2	1.5	1.4	1.1	0.48	0.48	0.48	0.48
DD	03z09	1.2	3.2	3.2	1.2	3.32	5.28	3.32	3.32
CS	03z10	2.2	5.5	3.5	4.1	4.04	4.04	6.1	15.8
ESCS	03z11	26.1	33.1	31.2	28.1	36.2	55.8	53.84	46.0
ESCS	03z12	28	36	35	26	30.32	32.28	34.24	40.12
SCS	03z13	5	5	1	1	11.88	11.88	11.88	5.76

The simulated total cloud fractions at 00 UTC on 12 October 2013 when Phailin reached ESCS stage are shown in Figure 6. The dense number of closed isobars (Figure 6a,c,e,g) following high clouds are represented in Figure 6b,d,f,h. The INSAT 3A satellite pictures (IMD, 2013) show heavy dense clouds aligned around the cyclone in Figure 6i. Similarly, the simulated circularly organized cloud bands with the extension of clouds in the northeast and southeast sectors indicate a broken comma structure. The cloud imager clearly shows a significant eyewall formation in MP experiments. The comparisons of simulated wind, pressure distribution, and cloud fractions from different experiments (MY, WSM6, WSM5, and WSM3) with NOAA and the INSAT 3A cloud imagery are represented in Figures 1a and 6i. The results indicate a more intensive cyclone in MY, WSM6, and WSM5 schemes than WSM3 (Figure 6h). The central eyewall formation was well-simulated in all MP schemes and finely represents the observed cloud bands. However, the WSM3 scheme could simulate the spatial distribution, as well as the position of the eyewall region, more comprehensively than other MP schemes.

All the simulations underestimated the pressure drop and maximum sustained wind at different stages, such as DD, SC, and SCS, of the storm. It attained maximum intensity at 06 UTC, 10 October 2013, lasted for more than 48 h, and persisted over the Bay of Bengal. The MP experiments underestimated the pressure intensity by nearly 18 hPa, thus resulting in weaker wind speed (knots) than the IMD observations. These simulations seem to have produced gradual deepening and mature phases of Phailin close to the timings of the observed storm. The significant underestimation in the pressure drop between the model and the observed, starting from the pre-deepening to weakening phases of the storm, was probably due to the cold start initialization of the WRF model.

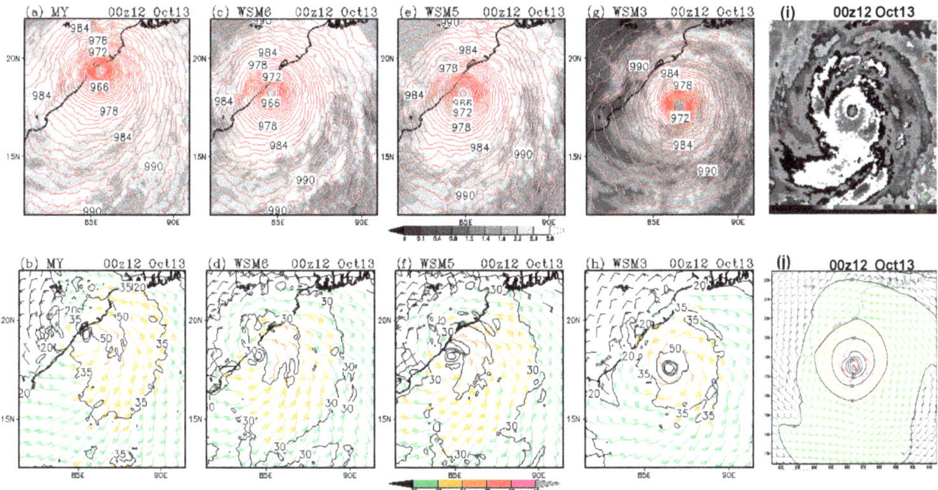

Figure 6. MSLP (hPa) and total cloud distributions (%) during the ESCS stage of Phailin derived from WRF simulation experiments using (**a**) MY, (**c**) WSM6, (**e**) WSM5, (**g**) WSM3 schemes, and (**i**) METOSAT 7 observation valid at 00 UTC, 12 October 2013 [initial condition of 00 UTC, on 8 October 2013] at 8.333 km resolutions; similarly, 10 m wind speed (shaded; knots) and direction obtained from (**b**) MY, (**d**) WSM6, (**f**) WSM5, (**h**) WSM3 schemes, and (**j**) CIRA observation respectively.

4.2. Circulations and Dynamical Mechanism for Eyewall Development

The movement of the cyclone is governed by the circulation and the intensification of eyewalls. Formation of the eye within any TC is one of the significant features owed to eyewall convection. The eyewall is created by organized convection that is lost for a longer period, with narrow rain bands, called spiral bands, oriented in the same direction as the horizontal wind speed appearing to spiral around the center of the TC [4]. In the case of Phailin, a significant eyewall was formed after 12 h of VSCS stage at 00 UTC, on 12 October 2013 [51–53]. Another important characteristic of the eyewall region is the warm temperatures (due to subsidence) that extend up to the tropospheric level, then to the surrounding environment, as discussed in [54]. Thus, the latent heat flux and temperature at the 300 hPa level are considered here. The contrast in temperature distributions between the warmest part of the eyewall and the coldest surrounding modulates the convection activity of VSCS Phailin (Figure 7i–l). The remarkable features, such as the core of the ESCS warmest of temperature (−14 °C) were simulated in all the schemes of the MY, WSM6, WSM5 and WSM3 experiments. Moreover, the WSM3 scheme showed the location of Phailin far away from the coastal Odisha, whereas MY, WSM6, and WSM5 simulated the location close to the coast. The curved band patterns of temperature, as depicted in Figure 6i–l, are associated with the lowest core of pressure and larger vorticity fields. The wind speed and direction at 10 m height, along with the averaged vertical velocity at 1000–700 hPa level derived from MP experiments, are shown in Figure 7e–h.

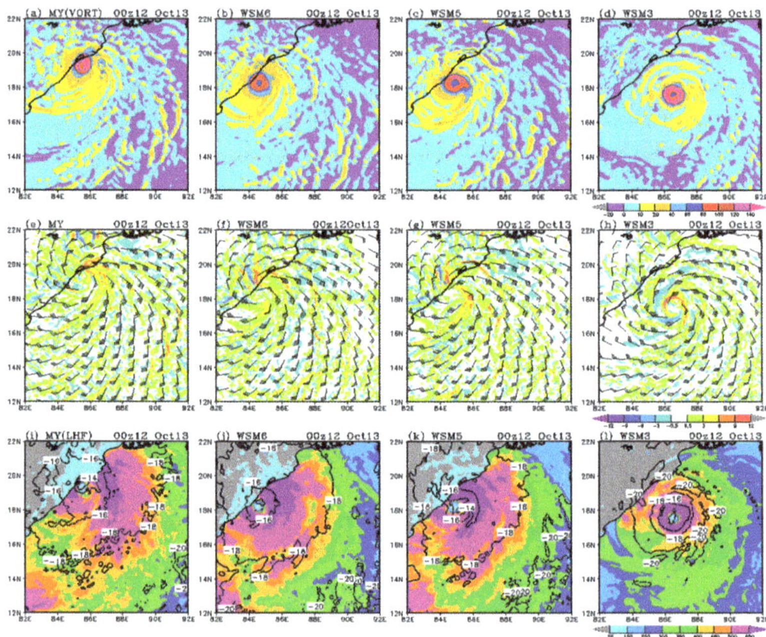

Figure 7. Average (1000–700 hPa level) vorticity field ($\times\ 10^5\ \mathrm{s}^{-1}$) derived from WRF (D02) experiments using (**a**) MY, (**b**) WSM6, (**c**) WSM5, and (**d**) WSM3 schemes, valid at 00 UTC, 12 October 2013; similarly, vertical velocity ($\times 10$; ms^{-1}) and 10 m wind fields (knots) obtained with (**e**) MY, (**f**) WSM6, (**g**) WSM5 and (**h**) WSM3 schemes are shown; along with latent heat flux (W m^{-2}, shaded) and temperature (°C; contours) at 300 hPa level derived from (**i**) MY, (**j**) WSM6, (**k**) WSM5, and (**l**) WSM3 experiments, respectively.

The surface winds are calm at the axis of rotation, while strong winds extend well into the eyewall of a TC, as reported by [55]. Interestingly, similar characteristics were captured by WRF simulations in the D02 region. The magnitude of 70 knots with a circular ring-like structure was captured by the WSM3 scheme (Figure 6h), whereas other microphysical parameterization schemes (MY, WSM6, and WSM5) were unable to represent the same (Figure 7e–g). However, except for WSM3, other MP processes have shown the fast movement of the vortex from the BoB towards the Odisha coast and that maximum wind intensity occurred in the east sector of ESCS Phailin. The MY scheme showed the location of the eyewall over the Chilka lagoon, which deviated a bit as compared to the satellite picture shown in Figure 1b. It is to be noted that the calm wind of 10 knots was well-captured by all of these MP schemes, but well-organized features were noticed in the WSM3 scheme.

The circulation budget was computed following the method employed by [56]. The rate of change of relative vorticity can be written in a form that relates to the circulation tendency within a boxed region, accounting for both eddy and mean contributions:

$$\frac{\partial C}{\partial t} = -\bar{\eta}\,\hat{\partial}A - \oint \acute{\eta}\, v \cdot n\ dl + \oint \omega\ \left(k \times \frac{\partial v}{\partial p} \right) \cdot n\ dl + \oint (k \times F) \cdot n\ dl \qquad (1)$$

where C is the circulation, η is the absolute vorticity, $\hat{\partial}$ is the mean divergence over the area A of the box, $\oint$ is the line integral around the perimeter of the box, v is the horizontal wind vector, n is the direction normal to the perimeter of the box, ω is the vertical velocity in the pressure coordinates, p is pressure, and F is the frictional force. To identify the key mechanisms responsible for the circular pattern of the eyewall, the average (1000–700 hPa) relative vorticity of Phailin was computed during the ESCS stage, valid at 00 UTC on 12 October 2013, and the same is shown in the upper panel of Figure 7a–d. The vorticity at 300 hPa (not shown) also confirmed the positive temperature anomalies over the upper level (300 hPa). Therefore, it supports and maintains circulation from surfaces to the upper level. The results included in Figure 7a–d reveal significant differences in the structure of Phailin. It seems both WSM3 and MY schemes were capable of simulating the well-organized circular structure of the relative vorticity field better than the WSM6 and WSM5 schemes. However, the latent heat fluxes, as shown in Figure 7i–l, were consequently favored with the relative vorticity budgets, and those that persisted over the BoB. MY and WSM3 schemes exhibited stronger relative vorticity than WSM5 and WSM6 schemes.

4.3. Middle and Vertical Atmospheric Features

Apart from the surface facilitating elements, vertical sustainability of temperature, water vapor, and wind, etc. are crucial. As Phailin gradually intensified, the location and intensity estimations became more accurate, with well-developed characteristic features of TCs, such as the eyewall, central dense overcast (CDO), lowest cloud top temperature (CTT), and curved band features [57]. Mohapatra et al. (2013) suggested that satellite and radar techniques are more appropriate for exact location and intensity forecasting of TCs over the Indian Ocean using [58]. Typically, the four types of spiral rainbands are principal, secondary, distant, and inner rain bands [59]. The principal rain band distributions were prominent in the case of ESCS Phailin. Therefore, horizontal cross-sections of equivalent potential temperature, relative humidity, and water vapor at 700 hPa (z = 3.02 km) at 00 UTC on 12 October 2013 are considered in this section. Figure 8 appears to fit the description of principal rain bands, as shown in Figure 1b. Figure 8a–h show equivalent potential temperature and relative humidity associated with the VSCS stage of Phailin. At 700 hPa (z = 3.02 km), a close-up view of principal rainbands in the WSM3 scheme (Figure 7d) produced more realistic features than other (MY, WSM6, and WSM 5) MP schemes. The performance of the WSM3 scheme for equivalent potential temperature (Figure 8d) and relative humidity (Figure 8h) showed significantly closer variations against observations, which may have been possible because of the slight increase in cloud ice and decrease of snow at warm temperatures during the microphysical process [57]. At 00 UTC, 12 October 2013, the spatial structure of relative humidity and water vapor mixing ratio coincided with the equivalent potential temperature. The MY, WSM6, and WSM5 schemes (Figure 8e–g) demonstrated convectively active elements of relative humidity (>70%) located close to the Odisha coast. In contrast, the center of Phailin in WSM3 reproduced over the oceanic region. Moreover, the MY scheme showed (Figure 8e) maximum intensity in terms of equivalent potential temperature, relative humidity, and water mixing ratio at 700 hPa level. Figure 9a–c is similar to Figure 8, but represents vertical profiles over Visakhapatnam meteorological station (17.68° N, 83.22° E) at 00 UTC, 12 October 2013, which can provide details on how MP schemes differ from each other from the surface to upper tropospheric level at a particular station. For all experiments, WSM3 was slightly different and better than the other MP schemes. The equivalent potential temperature value increased from the middle to upper tropospheric level (Figure 9a), whereas the water vapor mixing ratio decreased at all levels (Figure 9c). Three MP schemes, MY, WSM6, and WSM5, simulated the relative humidity above 90% of moisture contents from the surface up to 400 hPa level (Figure 9b), whereas WSM3 simulated lower than 90% of moisture content, which was closer to the observations. For this reason, it seems that WSM3 captured middle atmospheric features in a better way than other MP schemes.

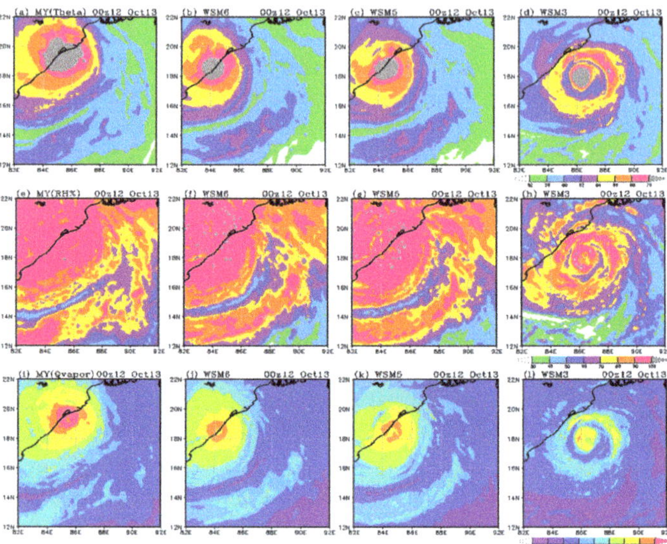

Figure 8. Model-simulated equivalent potential temperature (°C) at 700 hPa level derived from WRF (D02) experiments using (**a**) MY, (**b**) WSM6, (**c**) WSM5, and (**d**) WSM3 schemes valid at 00 UTC, 12 October, 2013. Figures (**e–l**) are the same as (**a–d**), but represent relative humidity (%) and water vapor mixing ratio ($\times 10^3$; gm·kg^{-1}) respectively.

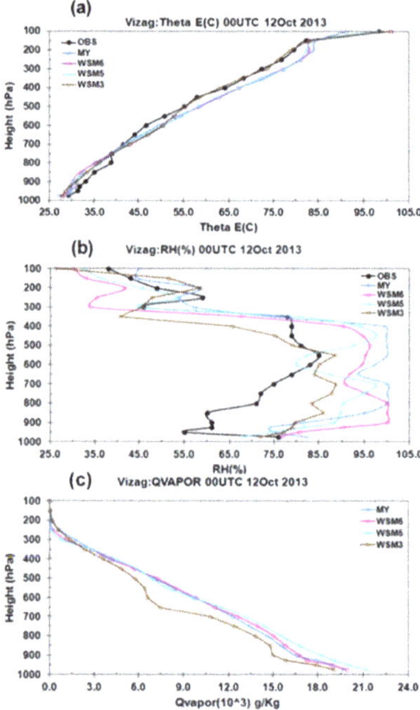

Figure 9. Vertical profiles of (**a**) equivalent potential temperature (°C), (**b**) relative humidity (%), and (**c**) water vapor mixing ratio (gmkg^{-1}) derived at Visakhapatnam meteorological station valid at 00 UTC, 12 October 2013.

The east–west vertical cross-section of temperature anomaly (°C), horizontal wind speed (ms^{-1}), and direction of Phailin at the ESCS stage valid at 00 UTC, 12 October 2013 are represented in Figure 10. The positive temperature anomaly indicates the shift of warming in the middle to upper tropospheric levels (600–150 hPa) during the ESCS stage of Phailin. The MY scheme simulated that maximum warming of 5 °C persisted above 500 to 200 hPa level (Figure 10a). However, an extra amount of warming was simulated in WSM6 (Figure 10c), WSM5 (Figure 9e), and WSM3 (Figure 10g) schemes over 400–300 hPa compared to the MY scheme.

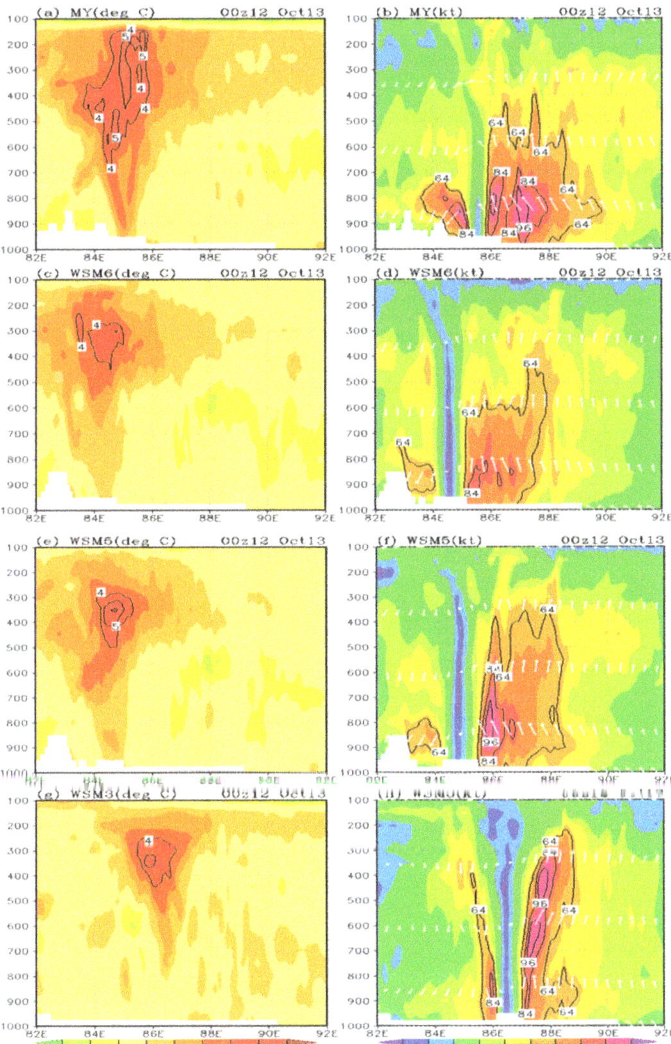

Figure 10. Vertical level and longitudinal cross-sections of (**a**) temperature anomaly (°C) and (**b**) wind field (shaded; ms^{-1}) derived from MY experiment; similarly (**c,d**), (**e,f**) and (**g,h**) were derived from WSM6, WSM5, and WSM3 experiments valid at 00 UTC, 12 October 2013 [as per Saffir–Simpson hurricane wind scale, the wind speed (contour) shown fits category 3, as simulated by WRF experiments].

This is consistent with earlier studies of Orissa super cyclone 1999 [59] and Andhra severe cyclone (2003), as discussed in Srinivas et al. (2007). One significant change was the temperature anomaly from the MP experiments, where the WSM3 single-moment schemed showed slower progress towards the Odisha coast than the double-moment schemes. The distribution of wind fields indicates that the presence of cyclonic winds remained in western (30–40 ms^{-1}) and eastern (maximum 45 ms^{-1}) sides, as noticed in the WSM3 experiment. In the vertical space, the calm wind also extended from the surface up to the upper tropospheric level, as captured by the WSM3 scheme. Moreover, WSM6 and WSM5 simulated maximum winds of 40 ms^{-1} up to 300 hPa level. In contrast, MY captured the maximum wind speed of 40 ms^{-1} limited up to 500 hPa level (Figure 10 b,d,f,h), which means that double moment MP cannot simulate the wind speed in the middle atmosphere during the ESCS stage of Phailin.

4.4. Rainfall Variability Due to Cyclone

A low-pressure system like a cyclone provides widespread and heavy rainfall over an extended region. One-day accumulated precipitation distribution associated with Phailin at the ESCS stage is illustrated in Figure 11. The heavy rain that occurred during the ESCS stage of Phailin according to TRMM and the rain predicted from the WRF model at 00 UTC, 12 October 2013 were almost similar in spatial pattern. Maximum precipitation bands occurred over the BoB, in the southwest zone of the center of Phailin during 00 UTC 12 October 2013.

Under the influence of ESCS, heavy to intense rainfall occurred over coastal Odisha, Andhra Pradesh, and adjoining areas. Critical analysis of rainfall from WRF simulation experiments and comparison against TRMM inferred that MP schemes such as MY, WSM6, WSM5, and WSM3 could capture the spatial pattern of rainfall correctly, but the WSM3 scheme simulation closely agreed with TRMM rainfall in terms of location and magnitude (Figure 11e). Besides, the rainfall from the IMD rain gauge stations of Odisha and Andhra Pradesh as valid at 03 UTC, on 12 October 2013 was comparable with the WRF experiments.

The magnitudes of rainfall derived from WRF model and IMD observations are included in Table 5. During landfall, it caused maximum damage to coasts and inland areas because the strong wind gust from the offshore region entered into the core of Phailin. The area average rainfall (mm·d^{-1}) over the domain of 82–92° E and 13–23° N during the simulation period (8–13 October 2013) is represented in Supplementary Figure S1. The 24, 48, 72, 96, and 120 h rainfall agreed well with the WSM3 scheme. However, 72, 96, and 120 h rainfall was substantially higher in MY, WSM6, and WSM5 schemes as compared with TRMM observation. Such differences are attributed to the inclusion of the different types of mixing ratios in microphysical parameterization schemes. Moreover, the rainfall distribution during the ESCS phase depends on the characteristics of hydrometeors that persisted in the Phailin storm. The magnitude of moisture convergence is also affected by the choice of cloud microphysical parameterization [60]. The spatial distribution of hydrometeors is analyzed and shown in Figures 12 and 13. The difference in the skills of microphysics schemes to simulate cloud and rainwater in the upper tropospheric levels will affect the total rainfall patterns. Moreover, WSM5 and WSM6 have shown that precipitating cloud and rainwater is more prominent within lower to intermediate levels, which may be the poor skill of these schemes.

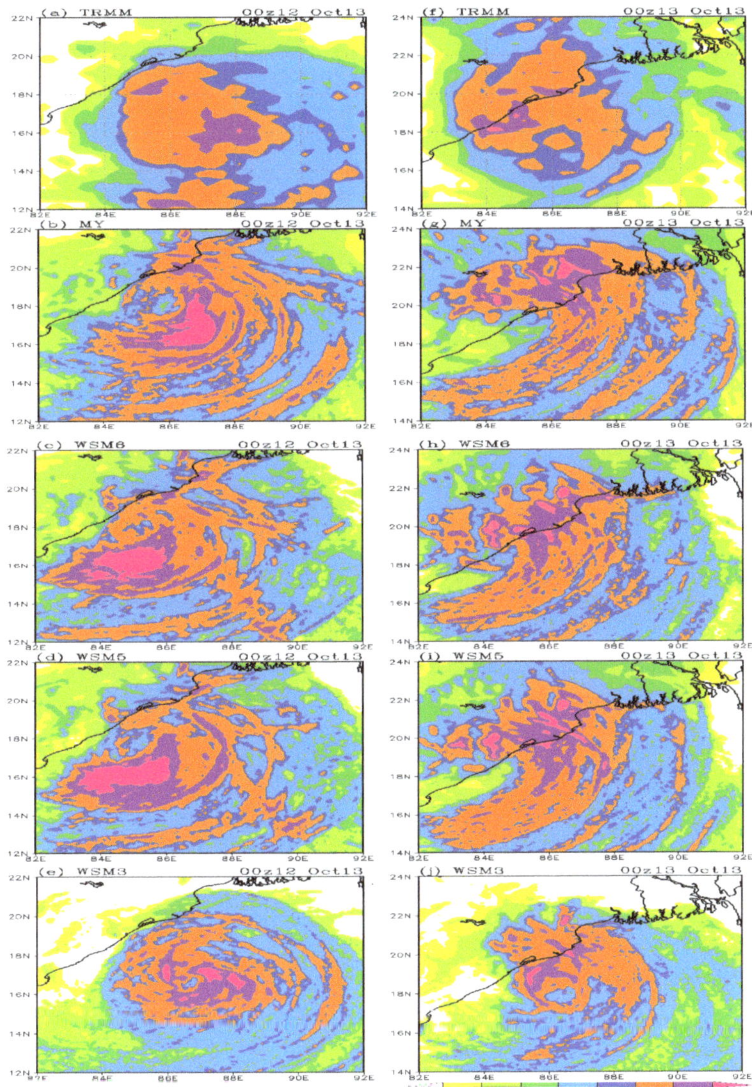

Figure 11. Comparison of 24 h accumulate precipitation (mm·d^{-1}) of Phailin at the ESCS stage derived from TRMM 3B42 satellite (**a**) and WRF model experiments using (**b**) MY, (**c**) WSM6, (**d**) WSM5, (**e**) WSM3 schemes, valid at 00 UTC, 12 October 2013 at D02 resolution; figures (**f–j**) are the same as (**a–e**), but valid at 00 UTC, 13 October 2013 respectively.

Table 5. Accumulated rainfall (cm) derived from IMD and WRF models (8.333 km) at D02 resolution at different stations over Odisha, coastal Andhra Pradesh, and Jharkhand, valid at 03 UTC, 12 October 2013.

State	Station	Latitude	Longitude	IMD (cm)	MY	WSM6	WSM5	WSM3
Orissa	Tikarpara	20.60	84.79	17	10.32	7.41	9.24	8.70
	Rajghat	21.07	86.50	92	14.94	12.34	16.42	8.05
	Nischintakoili	20.48	86.18	11	16.62	13.59	15.27	13.72
	Mundali	20.44	85.75	25	16.72	15.29	18.12	9.16
	Banki	20.38	85.53	38	13.19	12.25	15.67	6.51
	Hindol	20.61	85.20	23	13.72	12.23	14.04	9.93
	Mohana	19.44	84.26	19	11.96	18.62	19.48	11.56
	Ramba	19.51	85.09	14	1.54	11.22	11.14	8.23
	Purusottampur	19.52	84.89	18	1.82	8.15	9.82	4.01
	Chandikhol	20.71	86.10	15	14.08	19.32	23.78	15.17
	Danagadi	20.97	86.08	19	17.05	23.45	21.78	9.68
	Daringibadi	19.90	84.13	17	9.49	35.84	32.74	1.66
	Pattamundai	20.59	86.57	15	14.50	14.02	13.74	8.91
	Joda	22.02	85.41	19	14.03	12.01	11.81	4.56
	Banpur	19.77	85.16	20	5.71	18.63	14.55	8.79
	Bangiriposi	21.91	85.90	21	5.72	6.53	6.49	2.14
	Balimundali	21.74	86.63	31	22.40	23.74	18.83	16.19
	Nayagarh	20.12	85.10	18	9.27	11.27	13.99	10.86
	Ranpur	19.90	85.40	30	13.12	20.23	19.70	6.65
	Puri	19.81	85.83	12	5.64	16.64	12.32	8.05
Coastal AP	Palasa	18.76	84.42	10	0.49	5.78	3.92	3.80
	Sompeta	18.95	84.58	11	1.34	10.79	5.63	1.42
	Itchapuram	18.88	84.45	20	1.26	11.01	4.76	1.87
Jharkhand	Tenughat	23.73	85.79	7	3.58	2.58	4.38	1.20
	Dhanbad	23.80	86.43	7	3.15	2.52	3.65	0.50
	Chaibasa	22.55	85.80	20	13.2	5.28	8.17	2.62

The accurate simulation of different precipitation hydrometeors (shown in Table 3) leads to better rainfall prediction. Thus, WSM3 can capture the warm rainfall processes better than WSM5 and WSM6. The WSM3 scheme does not include mixed-phase microphysical processes, such as freezing, which occurs instantaneously where the temperature is colder than 0 °C, and melting, which occurs, similarly, one level below the freezing level. Moreover, the Phailin at the ESCS stage has a warm rain (>0 °C) process, thus representing the sub-grid-scale precipitations in a better way in WSM3 than MY, WSM6, and WSM5 schemes.

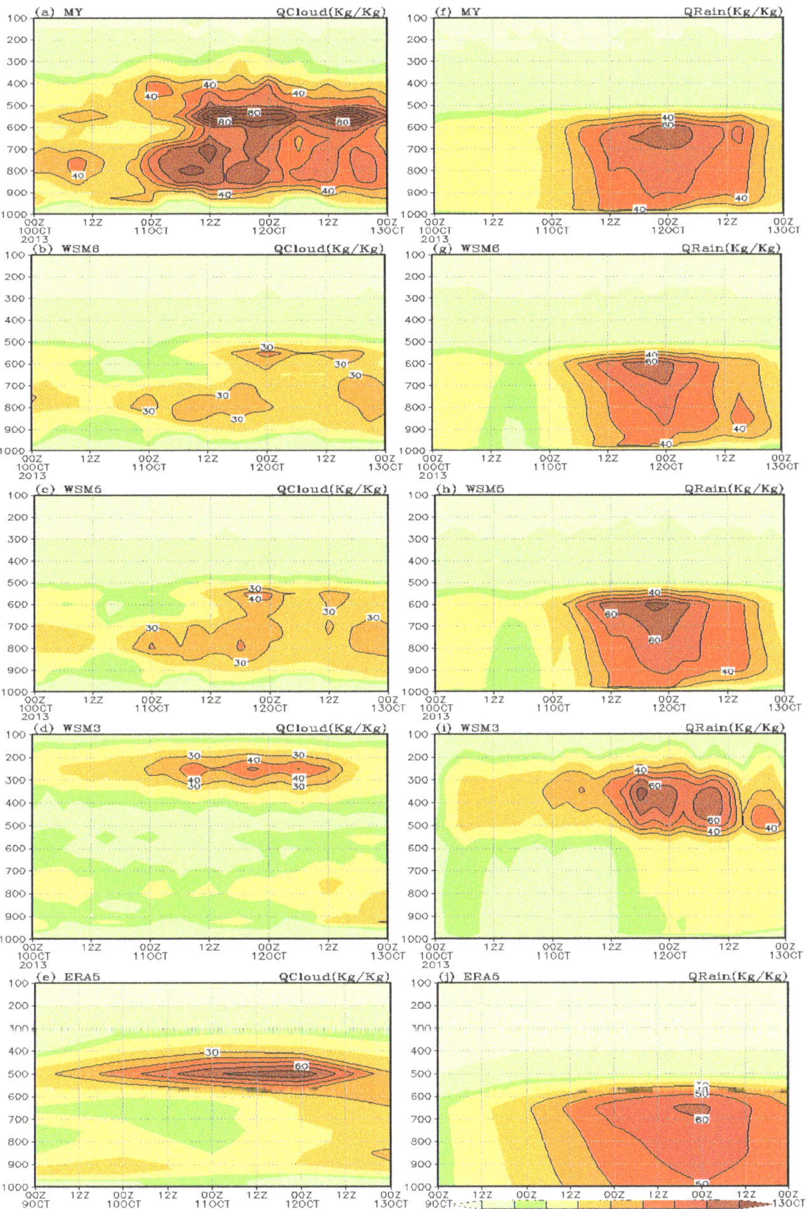

Figure 12. Time-pressure cross-section of the WRF simulations of Qcloud ($\times 10^6$ kg·kg^{-1}) derived from (**a**) MY (**b**) WSM6, (**c**) WSM5, (**d**) WSM3 schemes, and (**i**) ERA5 reanalysis with the averaged area over 82–92° E and 14–20° N from 00 UTC, from 10 to 13 October 2013, measured in 3-hourly intervals at D02 resolutions. Similarly, Qrain ($\times 10^6$ kg·kg^{-1}) mean mass contents are derived from (**e**) MY, (**f**) WSM6, (**g**) WSM5, and (**h**) WSM3 schemes, and (**j**) observation respectively.

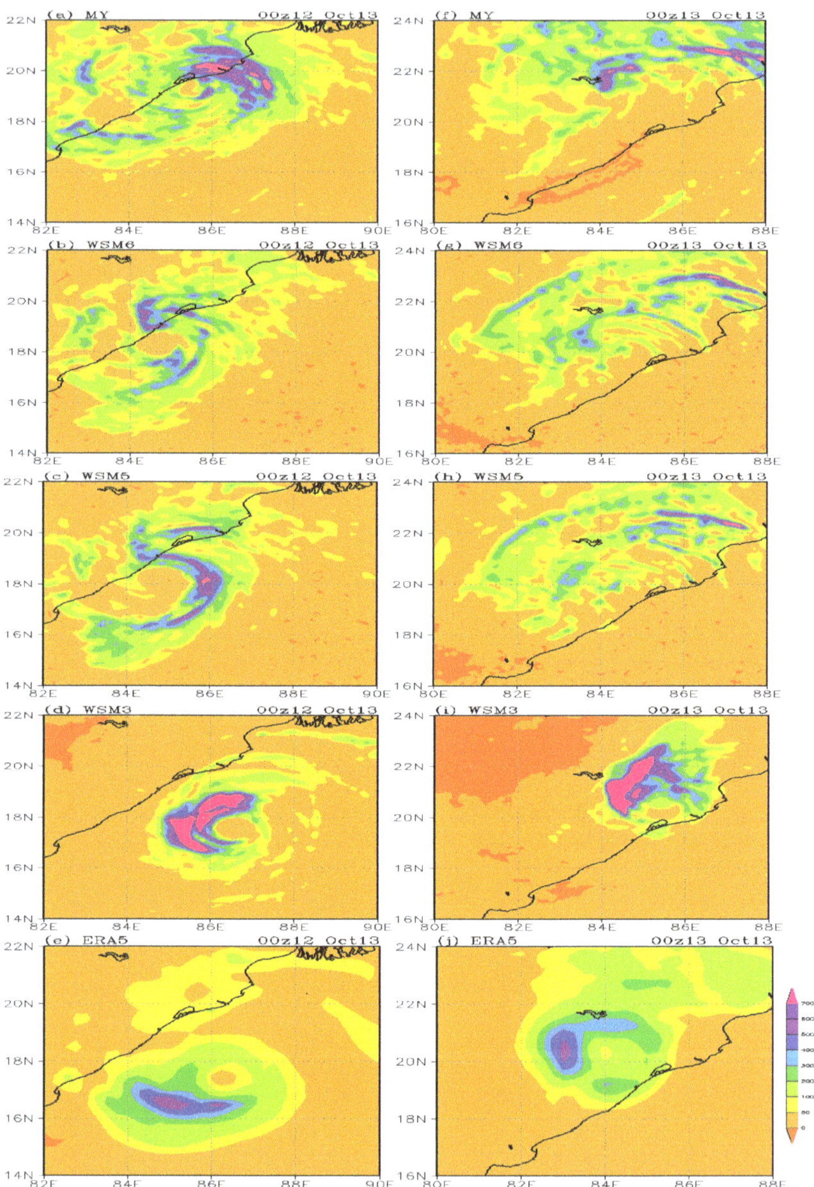

Figure 13. Horizontal distributions of average (1000–100 hPa) level precipitation hydrometeors (Qcloud and Qrain) derived from WRF (**a**) MY, (**b**) WSM6, (**c**) WSM5, and (**d**) WSM3 D02 experiments and (**e**) ERA5 reanalysis during the ESCS stage of Phailin valid at 00 UTC, 12 October 2013. Distributions in (**f–j**) are the same as (**a–e**), but valid at 00 UTC, 13 October 2013.

4.5. Characteristics of Precipitation Hydrometeors

Cossu and Hocke documented how the various microphysical processes in MP schemes are responsible for the differences in the rainfall-related variables [60]. Those represent various components of the precipitation (i.e., water cycle) [61] in the WRF model. The available MP schemes, ranging from simple, efficient, and sophisticated, are more computationally expensive. Moreover, both the newly developed schemes in the WRF model and well-used schemes, such as WSM3 and WSM6, are currently used in operational models [62]. Therefore, each scheme can simulate a certain number of variables, as discussed in Table 2. Consequently, the structures of spatial distributions of hydrometeors are significant in the case of ESCSs. Hence, this is further discussed in this section. Among the different water mixing ratios, Qcloud and Qrain are referred to as precipitation hydrometeors [39,63]. The precipitation hydrometeors, Qcloud and Qrain, are shown in Figure 12. The model grid points in the 8.333 km simulation region (D02) are recognized as the leading rainfall occurrence regions during ESCS phase of Phailin, as discussed in the previous section (Figure 10). The area averages of hydrometeors such as cloud water mixing ratio (Qcloud) and rainwater mixing ratio (Qrain) from00 UTC, from 9 to 13 October 2013 in a combination of four MP schemes are discussed. A time–height series of averaged hydrometeors, such as Qcloud and Qrain, are shown in Figure 12. The average was computed within area D02 where Phailin intensified (i.e., attained minimum MSLP). The amounts of the mean cloud water contents in the four experiments were significantly different from each other. The ESCS Phailin simulated with MY scheme produced maximum Qcloud water (>5 $gm \cdot kg^{-1}$) as compared to the other three experiments (WSM6, WSM5, and WSM3), and the peak values extended from 900 hPa to 300 hPa level during VSCS stages from 00 UTC, from 11–12 October 2013.

The experiments with WSM6 and WSM5 schemes generated similar Qcloud characteristics (900 hPa to 500 hPa), as shown in Figure 12c,e; however, WSM3 showed very different results than these schemes. The experiment with the WSM3 scheme showed two distinct features: (1) The maximum mixing ratios (5 $gm \cdot kg^{-1}$) were in between 300 hPa to 200 hPa levels and (2) another maximum was (2 $gm \cdot kg^{-1}$) at 800 hPa to 600 hPa level (Figure 12g). The evolution of the Qrain rate is examined in Figure 11e–h. The area-averaged mixing ratios were computed within the inner core region of ESCS Phailin. The convection processes produced the main features of the rainwater contents. The different hydrometeor distributions, such as rainwater contents, induced the different structures during the intensity phase of Phailin (11–12 October 2013). MP schemes MY, WSM6, and WSM5 showed relatively similar rainwater content patterns with slightly different intensities (Figure 12e–g). The maximum Qrain of 5 $gm \cdot kg^{-1}$ from the surface to 500 hPa level was simulated in MY, WSM6, and WSM5, whereas WSM3 captured the maximum peak shown in Figure 11h above 600 hPa to 150 hPa level. Overall, the vertical patterns of cloud and rainwater mixing ratios pointed out the other moisture representations in each MP scheme. The horizontal distributions of precipitation hydrometeors are measured by the combination of both Qcloud (non-precipitating hydrometers) and Qrain (Figure 13). The vertical level averages (1000–400 hPa) of the horizontal distribution of precipitation hydrometers derived from four MP schemes (MY, WSM6, WSM5, and WSM3), valid at 00 UTC on 12 October 2013, are shown in Figure 13a,c,e,g.

The horizontal distributions of precipitation hydrometeors showed a diminishing rate of precipitation hydrometeors with MY, WSM6, and WSM5 schemes. Again MY, WSM6, and WSM5 schemes generated a lesser amount of precipitation hydrometeors, both horizontally and vertically. However, hydrometeor precipitations were slow-moving towards the coast of Odisha, as captured in the WSM3 scheme, which was close to satellite images of Figure 1c,d.

5. Impact of Global Warning on Cyclones

Warming in the tropical Indian Ocean has increased faster, at 0.15 °C/decade during 1951–2015, compared to the global ocean at 0.11 °C/decade [64]. Researchers added that this warming has been non-uniform and ~90% of warming is attributed to anthropogenic activities. Changes in the SST and moisture availability (specifically humidity) for two decades have contributed positively to TC formation [65], as shown in Figure 14a,b. The warming over the western and central Indian Ocean is one of a few prominent features of local warming. The availability of moisture in the atmosphere in the recent decade is an essential aspect of the rapid intensification and strengthening of tropical cyclones before landfall (Figure 14b).

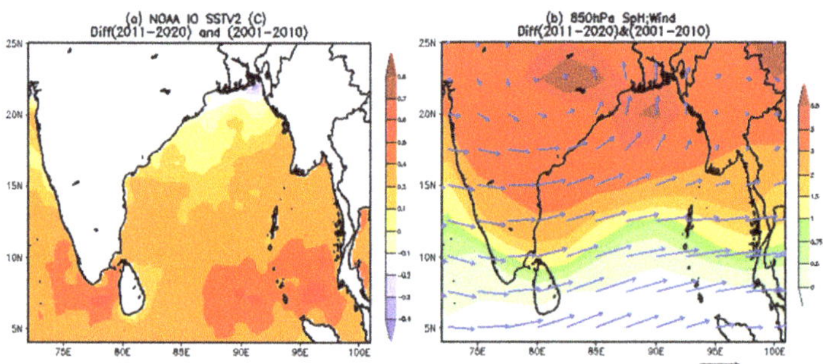

Figure 14. Variations of cyclone related variables over the BoB: (**a**) difference in SST (°C) and (**b**) specific humidity (shaded; gm·kg^{-1}), along with wind field (ms^{-1}) at 850 hPa level during two recent decades 2011–2020 and 2001–2010, derived from NOAA reanalysis.

SST over the North Indian Ocean (NIO) has been intensifying in recent decades due to global warming [66,67]. Based on the rising of ocean temperatures, consistent theory by Elsner (2020) has suggested that tropical cyclones during 1981–2006 and 2007–2019 showed the strongest tropical activity, which is getting stronger with time [68]. Over the Bay of Bengal basin, numerous studies have been conducted with an upward intensity trend of ESCSs. Under the global warming scenarios, there may be several plausible reasons for the rapid intensification of storms just before landfall. However, SST and moisture supply are two driving factors over the Indian Ocean during the post-monsoon season. Table 6 shows the formation of the total number of low-pressure systems over the Bay of Bengal and the Arabian Sea during October–November in the last 20 years (2000–2020). There were significant variations in the number of cyclones formed in the first decade 1991–2000. The depressions have decreased while severe cyclones have increased. Furthermore, due to warming, the intensity of the land-falling cyclones from severe to very severe cyclonic storms (SCS/VSCS) have escalated 24 h before landfall.

Table 6. Decadal variability in the number of tropical storms over the Bay of Bengal during the months of October–November, 2001–2020.

Year	DD	* SCS	ESCS	Duration of ESCS ≥ 24 h
2001–2010	20	14	2	2
2011–2020	19	11	6	6

* SCS = SC + SCS; DD = Deep Depression; SCS = Severe Cyclonic Storms.

The number of ESCSs has increased threefold during 2011–2020 as compared to 2001–2010 over the Bay of Bengal (Table 7). Such changes are attributed to the enhancement in the necessary and sufficient conditions for cyclone formations in the region, which includes the supply of abundant moisture and warmer SST in recent years (Figure 14). The comparison of SST, moisture, and winds in the last two decades shows the favorable conditions that facilitated the formation of VSCS.

Table 7. Different stages of tropical cyclones' mean sea level pressure (hPa) and maximum sustainable wind (knot) from October–November, during the 2011 to 2020 and 2001 to 2010 decades over the Bay of Bengal.

Stage	VSCS/ESCS during 2011–2020						VSCS/ESCS during 2001–2010	
	Phailin (MSLP/Wind)	Lahar (MSLP/Wind)	Hudhud (MSLP/Wind)	Titli (MSLP/Wind)	Gaja (MSLP/Wind)	Bulbul (MSLP/Wind)	Sidr (MSLP/Wind)	Giri (MSLP/Wind)
D	1004 (25)	1004 (25)	1004 (25)	1002 (25)	1002 (25)	1004 (45)	1004 (25)	1002 (25)
DD	1001 (30)	1002 (30)	1000 (30)	1000 (30)	1000 (30)	1001 (30)	1002 (30)	1002 (30)
SC	999 (35)	996 (45)	990 (45)	998 (35)	999 (35)	998 (35)	998 (40)	998 (50)
SCS	990 (55)	988 (55)	988 (60)	990 (55)	990 (55)	992 (50)	992 (55)	980 (60)
VSCS	984 (65)	982 (70)	970 (75)	972 (70)	988 (60)	983 (65)	986 (65)	976 (70)
ESCS	940 (115)	980 (75)	950 (100)	972 (80)	976 (70)	976 (75)	968 (90)	964 (90)
ESCS	940 (115)	988 (55)	950 (100)	972 (80)	-	982 (70)	944 (115)	950 (105)
SCS	990 (55)	998 (40)	987 (40)	996 (45)	999 (55)	998 (45)	1000 (45)	992 (45)
DD	996 (30)	1000 (30)	998 (30)	1001 (30)	1003 (30)	1002 (30)	1002 (25)	998 (25)

Table 7 represents the details of VSCS cyclone formations in the last two decades. There has been a notable shift in the total number of severe to extremely severe cyclones over the Bay of Bengal during the last 20 years. Regional or remote influence may provide a further explanation for such a shift. There may be a decadal or large-scale shift in the basic components of atmospheric and ocean conditions that facilitate the tropical cyclone formation. An increase in cyclonic activity over the BoB, due to SST-induced moisture supply, has been found. Studies by [69,70] have suggested that aerosol concentrations in the middle atmosphere play an important role in latent heat release that probably reduces the basin-wide vertical wind shear and creates a favorable environment for intense tropical cyclones over the NIO. Other methods are explained in [71] to understand the role of anthropogenic influence on cyclones. The impact of local, regional, and global warming on cyclone intensification and formation will have an impact on property damage, affecting buildings and flood extents.

6. Conclusions

This study intended to address the role of the microphysical processes influencing the simulations from the initial phase to the ESCS stage of Phailin over the BoB. The fundamental difference between two MP schemes, double- and single-moment, is illustrated at 8.333 km of the resolution, a gray-zone simulation. To this end, we analyzed several parameters, such as MSLP, 10m wind, track, cloud, precipitation, and hydrometeors, to study the differences in the simulated track, intensity, and structural evolution of Phailin. The key findings are highlighted below.

- The gray-zone simulations of track, intensity, and precipitation processes during ESCS stage were sensitive to the parameterization of different microphysical processes.
- One of the most sensitive results found that the eyewall clouds and characteristics during the deep depression (DD) to severe cyclonic storm (SCS) stage were better represented by the MY scheme than other MP schemes. However, WSM3 showed relatively better results regarding landfall over the Odisha coast and lower track errors than other MP schemes (Figure 4).
- Above 700 hPa, the water vapor in the cloud condenses into water, droplets releasing the latent heat, which originally evaporates the water. Latent heat provides the energy to drive the tropical cyclone circulation. However, lower heat release was utilized by Phailin to lower its surface pressure and increase the wind speeds (Figure 6).
- The eyewall development and its dynamical mechanism were analyzed with double- and single-moment MP schemes, which are sensitive and indicate the linkage between water vapor on one side and precipitation on the other. All the MP schemes simulated rainfall in terms of location and magnitude and closely agreed with TRMM rainfall (Figure 10e).

- Based on the above results, we can conclude that a double-moment cloud microphysical scheme (MY) is preferred for cyclonic systems. However, the MY scheme is unable to simulate the wind speed in the middle atmosphere.
- The clouds associated with Phailin that contribute to the vertical and horizontal redistribution of water vapor contents are well-simulated in all the schemes (WSM3, MY, WSM6, and WSM5).
- It is also important to assess the sensitivity of the simulated results of the double-moment MP schemes in WRF to provide helpful information towards improving cloud microphysics parameterization in the future. All the schemes show different results. Thus, a unified scheme is suggested for a consensus among various schemes by applying equal/unequal weight.
- There is a shift in cyclone formation over the Indian Ocean due to regional warming and availability of moisture supply in the Bay of Bengal, which has been especially evident over the western region in the recent decade (2011–2020) compared to the previous decade (2001–2010).
- The uncertainty that arises due to model physics could be identified with individual simulations, but multi-physics ensemble techniques using a number of physical parameterization schemes (PBL, cumulus convection, and cloud microphysics) are better at simulating track and intensity of TCs over NIO.

We are not certain if such a shift in the formation of cyclones over the Indian Ocean basin is temporary or permanent. In forthcoming articles, we will use the simulation of 500 m resolution datasets of multiple ESCSs to further focus on microphysical schemes to simulate more TCs (processes in the formation of convective cells) over the BoB and the Arabian Sea. Variability in the cyclone formation and rapid intensity of TCs just before landfall during the post-monsoon season (October–December) can be studied.

Supplementary Materials: The following are available online at https://www.mdpi.com/article/10.3390/oceans2030037/s1, Figure S1: The time-series of simulated rainfall (mmd^{-1}) averaged over the domain (82–92° E and 13–23° N) derived from observation and MY, WSM6, WSM5 and WSM3 experiments during 24-, 48-, 72-, 96- and 120 h).

Author Contributions: P.K.P. and V.K. conceived and formulated the study. P.K.P., V.K., S.K. and T.S. prepared the Figures. All other authors (S.V.B.R., V.K.K. and S.P.) along with P.K.P. discussed the results and prepared the manuscript. All authors have read and agreed to the published version of the manuscript.

Funding: This research received no external funding.

Institutional Review Board Statement: We used WRF model simulation done by first author, ERA-Interim reanalysis and observational datasets. These datasets are public.

Informed Consent Statement: Not applicable.

Data Availability Statement: The datasets will be available to anyone on request.

Acknowledgments: The authors (P.K.P. and V.K.K.) are thankful to ISRO-RESPOND Project, Govt. of India, for providing financial and necessary facilities to carry out this research work. We are also grateful to NCAR/NCEP team members for providing the ARW-WRF model and initial and boundary conditions to conduct the simulation experiments. IMD is sincerely acknowledged for providing the observations (best track, wind speed, and Kalpana satellite products) for model validations. In addition, the authors acknowledge the TRMM, Cooperative Institute for Research in the Atmosphere (CIRA) and NOAA satellite pictures that were used for validation. The figures prepared using GraDS software is sincerely acknowledged. We would like to thank the anonymous reviewers for their constructive comments and suggestions which significantly helped to improve this manuscript.

Conflicts of Interest: The authors declare no conflict of interest.

References

1. Kumar, T.S.V.V.; Sanjay, J.; Basu, B.K.; Mitra, A.K.; Rao, D.V.B.; Sharma, O.P.; Pal, P.K.; Krishnamurti, T.N. Experimental superensemble forecasts of tropical cyclones over the Bay of Bengal. *Nat. Hazards* **2006**, *41*, 471–485. [CrossRef]
2. Bhaskar Rao, D.V.; Srinivas, D. Multi-Physics ensemble prediction of tropical cyclone movement over Bay of Bengal. *Nat. Hazards* **2014**, *70*, 883–902.
3. Jullien, S.; Marchesiello, P.; Menkes, C.; Lefèvre, J.; Jourdain, N.; Samson, G.; Lengaigne, M. Ocean feedback to tropical cyclones: Climatology and processes. *Clim. Dyn.* **2014**, *43*, 2831–2854. [CrossRef]
4. Mohapatra, M.; Sharma, M.; Devi, S.S.; Kumar, S.V.J.; Sabade, B.S. Frequency of genesis and landfall of different categories of tropical cyclones over the North Indian. *Ocean. Mausam* **2021**, *72*, 1–26.
5. Kumar, V.; Sahu, D.K.; Simon, A.; Thomas, A.; Sinha, T.; Bielli, J.; Krishnamurti, T.N. Buoyancy Streams and Cloud Flare-Ups Along Rainbands in the Eye Wall of Rapidly Intensifying Storms. *Pure Appl. Geophys. PAGEOPH* **2021**, *178*, 223–243. [CrossRef]
6. Rao, D.V.B.; Prasad, D.H.; Srinivas, D. Impact of horizontal resolution and the advantages of the nested domains approach in the prediction of tropical cyclone intensification and movement. *J. Geophys. Res. Space Phys.* **2009**, *114*. [CrossRef]
7. Houze, R.A. Clouds in Tropical Cyclones. *Mon. Weather. Rev.* **2010**, *138*, 293–344. [CrossRef]
8. Pattnaik, S.; Inglish, C.; Krishnamurti, T.N. Influence of Rain-Rate Initialization, Cloud Microphysics, and Cloud Torques on Hurricane Intensity. *Mon. Weather. Rev.* **2011**, *139*, 627–649. [CrossRef]
9. Bao, J.-W.; Michelson, S.A.; Grell, E.D. Pathways to the Production of Precipitating Hydrometeors and Tropical Cyclone Development. *Mon. Weather. Rev.* **2016**, *144*, 2395–2420. [CrossRef]
10. Stephens, G.L. Cloud feedbacks in the climate system: A critical review. *J. Clim.* **2005**, *18*, 237–273. [CrossRef]
11. Randall, D.; Khairoutdinov, M.; Arakawa, A. Grabowski W Breaking the cloud parameterization dead-lock. *Bull Am. Meteorol. Soc.* **2003**, *84*, 1547–1564. [CrossRef]
12. Grabowski, W.W.; Morrison, H.; Shima, S.I.; Abade, G.C.; Dziekan, P.; Pawlowska, H. Modeling of cloud mi-crophysics: Can we do better? *Bull Am. Meteorol. Soc.* **2019**, *100*, 655–672. [CrossRef]
13. Hong, S.Y.; Dudhia, J.; Chen, S.H. A revised approach to ice microphysical processes for the bulk parameterization of clouds and precipitation. *Mon. Weather Rev.* **2004**, *132*, 103–120. [CrossRef]
14. Taraphdar, S.; Pauluis, O.M.; Xue, L.; Liu, C.; Rasmussen, R.; Ajayamohan, R.S.; Tessendorf, S.; Jing, X.; Chen, S.; Grabowski, W.W. WRF Gray-Zone Simulations of Precipitation Over the Middle-East and the UAE: Impacts of Physical Parameterizations and Resolution. *J. Geophys. Res. Atmos.* **2021**, *126*, e2021JD034648. [CrossRef]
15. Srinivas, C.V.; Yesubabu, V.; Venkatesan, R.; Ramarkrishna, S.S.V.S. Impact of assimilation of conventional and satellite meteoro-logical observations on the numerical simulation of a Bay of Bengal Tropical Cyclone of November 2008 near Tamil Nadu using WRF model. *Meteorol. Atmos. Phy.* **2009**, *110*, 19–44. [CrossRef]
16. Osuri, K.K.; Mohanty, U.C.; Routray, A.; Kulkarni, M.A.; Mohapatra, M. Customization of WRF-ARW model with physical parameterization schemes for the simulation of tropical cyclones over North Indian Ocean. *Nat. Hazards* **2012**, *63*, 1337–1359. [CrossRef]
17. Skamarock, W.C.; Klemp, J.B.; Dudhia, J.; Gill, D.O.; Barker, D.M.; Dudha, M.G.; Huang, X.; Wang, W.; Powers, J.G. A Description of the Advanced Research WRF Version 3. NCAR Technical Note NCAR/ TN-475? STR. 2008, p. 113. Available online: http://www2.mmm.ucar.edu/wrf/users/docs/arw_v3.pdf (accessed on 21 August 2021).
18. Srinivas, C.V.; Bhakar Rao, D.V.; Yesubabu, V.; Baskarana, R.; Venkatraman, B. Tropical Cyclone Predic-tions over the Bay of Bengal Using the High Resolution Advanced Research Weather Research and Fore-casting (ARW) Model. *Q. J. R Meteorol. Soc.* **2012**, *139*, 1810–1825. [CrossRef]
19. Mohanty, U.C.; Osuri, K.K.; Tallapragada, V.; Marks, F.; Pattanayak, S.; Mohapatra, M.; Rathore, L.S.; Gopalakrishnan, S.G.; Niyogi, D. A Great Escape from the Bay of Bengal "Super Sapphire–Phailin" Tropical Cyclone: A Case of Improved Weather Forecast and Societal Response for Disaster Mitigation. *Earth Interact.* **2015**, *19*, 1–11. [CrossRef]
20. Bhattacharya, S.K.; Kotal, S.D.; Roy Bhowmik, S.K.; Kundu, P.K. Sensitivity of WRF-ARW Model to Cumu-lus Parameterisation Schemes in Prediction of TC Intensity and Track Over the North Indian Ocean. In *Tropical Cyclone Activity over the North Indian Ocean*; Springer: Cham, Switzerland, 2017; pp. 295–306.
21. Choudhury, D.; Das, S. The sensitivity to the microphysical schemes on the skill of forecasting the track and Intensity of tropical cyclones using WRF-ARW model. *J. Earth Syst. Sci.* **2017**, *126*, 57. [CrossRef]
22. Krishnamurti, T.N.; Dubey, S.; Kumar, V.; Deepa, R.; Bhardwaj, A. Scale interaction during an extreme rain event over southeast India. *Q. J. R. Meteorol. Soc.* **2017**, *143*, 1442–1458. [CrossRef]
23. Gallus, W.A.; Pfeifer, M. Intercomparison of simulations using 5 WRF microphysical schemes with dual-Polarization data for a German squall line. *Adv. Geosci.* **2008**, *16*, 109–116. [CrossRef]
24. Reeves, H.D.; Ii, D.D. The Dependence of QPF on the Choice of Microphysical Parameterization for Lake-Effect Snowstorms. *J. Appl. Meteorol. Clim.* **2013**, *52*, 363–377. [CrossRef]
25. Wu, L.; Wang, B. Effects of Convective Heating on Movement and Vertical Coupling of Tropical Cyclones: A Numerical Study. *J. Atmospheric. Sci.* **2001**, *58*, 3639–3649. [CrossRef]
26. Wang, Y. How Do Outer Spiral Rainbands Affect Tropical Cyclone Structure and Intensity? *J. Atmospheric Sci.* **2009**, *66*, 1250–1273. [CrossRef]

27. Mohanty, U.C.; Mandal, M.; Raman, S. Simulation of Orissa Super Cyclone using PSU/NCAR Mesoscale Model. *Nat. Hazards* **2004**, *31*, 373–390. [CrossRef]
28. Rao, D.V.B.; Prasad, D.H. Sensitivity of tropical cyclone intensification to boundary layer and convective processes. *Nat. Hazards* **2006**, *41*, 429–445. [CrossRef]
29. Mahala, B.K. Pratap Kumar Mohanty, Birendra Kumar Nayak Impact of Microphysics Schemes in the Simulation of Cyclone Phailin using WRF model. *Procedia Eng.* **2015**, *116*, 655–662. [CrossRef]
30. Mandal, M.; Singh, K.S.; Balaji, M.; Mohapatra, M. Performance of WRF-ARW model in real-time pre-diction of Bay of Bengal cyclone 'Phailin'. *Pure Appl. Geophys.* **2016**, *173*, 1783–1801. [CrossRef]
31. Kumar, S.; Routray, A.; Tiwari, G.; Chauhan, R.; Jain, I. Simulation of Tropical Cyclone 'Phailin' Using WRF Modeling System. In *Tropical Cyclone Activity over the North Indian Ocean*; Springer International Publishing: Berlin/Heidelberg, Germany, 2017; pp. 307–316.
32. Hong, S.Y.; Noh, Y.; Dudhia, J. A new vertical diffusion package with an explicit treatment of entrain-ment processes. *Mon. Wea. Rev.* **2006**, *134*, 2318–2341. [CrossRef]
33. Grell, G.A.; Dévényi, D. A generalized approach to parameterizing convection combining ensemble and data assimilation techniques. *Geophys. Res. Lett.* **2002**, *29*, 38-1–38-4. [CrossRef]
34. Pattanayak, S.; Mohanty, U.C.; Osuri, K.K. Impact of Parameterization of Physical Processes on Simulation of Track and Intensity of Tropical Cyclone Nargis with WRF-NMM Model. *Sci. World J.* **2012**, *2012*, 1–18. [CrossRef]
35. Sandeep, C.P.; Krishnamoorthy, C.; Balaji, C. Impact of cloud parameterization schemes on the simula-tion of cyclone Vardah using the WRF model. *Cur. Sci.* **2018**, *115*, 1143–1153. [CrossRef]
36. Dudhia, J. A multi-layer soil temperature model for MM5. In Proceedings of the Sixth PSU/NCAR Mesoscale Model Users Workshop, Boulder, CO, USA, 22 July 1996; pp. 22–24.
37. Dudhia, J. Numerical study of convection observed during the winter monsoon experiment using a mesoscale two dimensional model. *J. Atmos Sci.* **1989**, *46*, 3077–3107. [CrossRef]
38. Huang, Y.J.; Cui, X.P.; Wang, Y.P. Cloud microphysical differences with precipitation intensity in a tor-rential rainfall event in Sichuan, China. *Atmos Oceanic. Sci. Lett.* **2016**, *9*, 90–98. [CrossRef]
39. McCumber, M.; Tao, W.K.; Simpson, J.; Penc, R.; Soong, S.T. Comparison of ice-phase microphysical pa-rameterization schemes using numerical simulations of tropical convection. *J. App. Meteorol.* **1991**, *30*, 985–1004. [CrossRef]
40. Bony, S.; Stevens, B.; Frierson, D.M.W.; Jakob, C.; Kageyama, M.; Pincus, R.; Shepherd, T.G.; Sherwood, S.C.; Siebesma, A.P.; Sobel, A.; et al. Clouds, circulation and climate sensitivity. *Nat. Geosci.* **2015**, *8*, 261–268. [CrossRef]
41. Morrison, H.; Milbrandt, J.A. Parameterization of Cloud Microphysics Based on the Prediction of Bulk Ice Particle Properties. Part I: Scheme Description and Idealized Tests. *J. Atmospheric Sci.* **2015**, *72*, 287–311. [CrossRef]
42. Jensen, E.J.; Pfister, L.; Jordan, D.E.; Bui, T.V.; Ueyama, R.; Singh, H.B.; Thornberry, T.D.; Rollins, A.W.; Gao, R.-S.; Fahey, D.W.; et al. The NASA Airborne Tropical Tropopause Experiment: High-Altitude Aircraft Measurements in the Tropical Western Pacific. *Bull. Am. Meteorol. Soc.* **2017**, *98*, 129–143. [CrossRef]
43. Michalakes, J.; Chen, S.; Dudhia, J.; Hart, L.; Klemp, J.; Middlecoff, J.; Skamarock, W. Development of a Next-Generation Regional Weather Research and Forecast Model. In *Developments in Teracomputing*; World Scientific: Singapore, 2001; pp. 269–296.
44. Chen, S.-H.; Sun, W.-Y. A One-dimensional Time Dependent Cloud Model. *J. Meteorol. Soc. Jpn.* **2002**, *80*, 99–118. [CrossRef]
45. Hong, S.-Y.; Pan, H.-L. Convective Trigger Function for a Mass-Flux Cumulus Parameterization Scheme. *Mon. Weather. Rev.* **1998**, *126*, 2599–2620. [CrossRef]
46. Kessler, E. On the distribution and continuity of water substance in atmospheric circulations. *Meteor. Monogr.* **1969**, *10*, 88.
47. Hong, S.Y.; Lim, J.O. The WRF single-moment 6-class microphysics scheme (WSM6). *Asia-Pacific J. Atmos. Sci.* **2006**, *42*, 129–151.
48. Milbrandt, J.A.; Yau, M.K. A Multimoment Bulk Microphysics Parameterization. Part I: Analysis of the Role of the Spectral Shape Parameter. *J. Atmospheric Sci.* **2005**, *62*, 3051–3064. [CrossRef]
49. Milbrandt, J.A.; Yau, M.K. A Multimoment Bulk Microphysics Parameterization. Part II: A Proposed Three-Moment Closure and Scheme Description. *J. Atmospheric Sci.* **2005**, *62*, 3065–3081. [CrossRef]
50. IMD. *Very Severe Cyclonic Storm PHAILIN over the Bay of Bengal (08–14 October 2013): A Report*; India Meteorological Department (Ministry of Earth Sciences), Government of India: Delhi, India, 2013.
51. Chandrasekar, R.; Balaji, C. Sensitivity of tropical cyclone Jal simulations to physics parameteriza-tions. *J. Earth System. Sci.* **2012**, *121*, 923–946. [CrossRef]
52. Durai, V.R.; Kotal, S.D.; Bhowmik, S.R.; Bharadwaj, R. Performance of Global Forecast System for the Pre-diction of Intensity and Track of Very Severe Cyclonic Storm 'Phailin'over North Indian Ocean. In *High-Impact Weather Events over the SAARC Region*; Springer International Publishing: Berlin/Heidelberg, Germany, 2015; pp. 191–203.
53. Weatherford, C.; Gray, W.M. Typhoon structure as revealed by aircraft reconnaissance. Part II: Struc-tural variability. *Mon. Wea. Rev.* **1988**, *116*, 1044–1056. [CrossRef]
54. Hawkins, H.F.; Rubsam, D.T. Hurricane Hilda, 1964, II Structure and budgets of the hurricane on 01 Oct, 1964. *Mon. Wea. Rev.* **1968**, *104*, 418–442. [CrossRef]
55. Davis, C.A.; Galarneau, T.J., Jr. The vertical structure of mesoscale convective vortices. *J. Atmosp. Sci.* **2009**, *66*, 686–704. [CrossRef]
56. IMD. *Cyclone Manual*; IMD: Lodi Road, New Delhi, India, 2003.

57. Mohapatra, M.; Bandyopadhyay, B.K.; Nayak, D.P. Evaluation of operational tropical cyclone intensity forecasts over north Indian Ocean issued by India Meteorological Department. *Nat. Hazards* **2013**, *68*, 433–451. [CrossRef]
58. Moon, Y.; Nolan, D.S. Spiral Rainbands in a Numerical Simulation of Hurricane Bill. Part II: Propagation of Inner Rainbands. *J. Atmospheric Sci.* **2015**, *72*, 191–215. [CrossRef]
59. Hazra, V.; Pattnaik, S.; Sisodiya, A.; Baisya, H.; Turner, A.; Bhat, G. Assessing the performance of cloud microphysical parameterization over the Indian region: Simulation of monsoon depressions and validation with INCOMPASS observations. *Atmos. Res.* **2020**, *239*, 104925. [CrossRef]
60. Cossu, F.; Hocke, K. Influence of microphysical schemes on atmospheric water in the Weather Research and Forecasting model. *Geosci. Model Dev.* **2014**, *7*, 147–160. [CrossRef]
61. Wang, W.; Bruyère, C.; Duda, M.; Dudhia, J.; Gill, D.; Lin, H.C.; Michalakes, J.; Rizvi, S.; Zhang, X. *Weather Research & Forecasting ARW Version 3 Modeling System User's Guide, Mesoscale & Microscale Meteorology Division*; NCAR: Boulder, CO, USA, 2012.
62. Cui, X.; Wang, Y.; Yu, H. Microphysical differences with rainfall intensity in severe tropical storm Bilis. *Atm. Sci. Lett.* **2015**, *16*, 27–31. [CrossRef]
63. Roxy, M.K.; Gnanaseelan, C.; Parekh, A.; Chowdary, J.S.; Singh, S.; Modi, A.; Kakatkar, R.; Mohapatra, S.; Dhara, C.; Shenoi, S.C.; et al. Indian Ocean Warming. In *Assessment of Climate Change over the Indian Region*; Springer Science and Business Media LLC: Berlin/Heidelberg, Germany, 2020; pp. 191–206.
64. Li, Z.; Yu, W.; Li, K.; Wang, H.; Liu, Y.; Liu, L. Environmental Conditions Modulating Tropical Cyclone Formation over the Bay of Bengal during the Pre-Monsoon Transition Period. *J. Clim.* **2019**, *32*, 4387–4394. [CrossRef]
65. Lackmann, G.M. Hurricane Sandy before 1900 and after 2100. *Bull Am. Meteorol. Soc.* **2015**, *96*, 547–560. [CrossRef]
66. Cho, C.; Li, R.; Wang, S.-Y.; Yoon, J.-H.; Gillies, R.R. Anthropogenic footprint of climate change in the June 2013 northern India flood. *Clim. Dyn.* **2016**, *46*, 797–805. [CrossRef]
67. Villarini, G.; Zhang, W.; Quintero, F.; Krajewski, W.F.; A Vecchi, G. Attribution of the impacts of the 2008 flooding in Cedar Rapids (Iowa) to anthropogenic forcing. *Environ. Res. Lett.* **2020**, *15*, 114057. [CrossRef]
68. Elsner, J.B. Continued Increases in the Intensity of Strong Tropical Cyclones. *Bull. Am. Meteorol. Soc.* **2020**, *101*, E1301–E1303. [CrossRef]
69. Evan, A.T.; Kossin, J.; Chung, C.; Ramanathan, V. Arabian Sea tropical cyclones intensified by emissions of black carbon and other aerosols. *Nat. Cell Biol.* **2011**, *479*, 94–97. [CrossRef]
70. Rajeevan, M.; Srinivasan, J.; Niranjan Kumar, K.; Gnanaseelan, C.; Ali, M.M. On the epochal var-iation of intensity of tropical cyclones in the Arabian Sea. *Atmos. Sci. Lett.* **2013**, *14*, 249–255. [CrossRef]
71. Hazra, A.; Mukhopadhyay, P.; Taraphdar, S.; Chen, J.-P.; Cotton, W.R. Impact of aerosols on tropical cyclones: An investigation using convection-permitting model simulation. *J. Geophys. Res. Atmos.* **2013**, *118*, 7157–7168. [CrossRef]

Article

Simulated Changes in Tropical Cyclone Size, Accumulated Cyclone Energy and Power Dissipation Index in a Warmer Climate

Michael Wehner

Lawrence Berkeley National Laboratory, Berkeley, CA 94720, USA; mfwehner@lbl.gov

Abstract: Detection, attribution and projection of changes in tropical cyclone intensity statistics are made difficult from the potentially decreasing overall storm frequency combined with increases in the peak winds of the most intense storms as the climate warms. Multi-decadal simulations of stabilized climate scenarios from a high-resolution tropical cyclone permitting atmospheric general circulation model are used to examine simulated global changes from warmer temperatures, if any, in estimates of tropical cyclone size, accumulated cyclonic energy and power dissipation index. Changes in these metrics are found to be complicated functions of storm categorization and global averages of them are unlikely to easily reveal the impact of climate change on future tropical cyclone intensity statistics.

Keywords: tropical cyclones; climate change; accumulated cyclone energy index; power dissipation index

Citation: Wehner, M. Simulated Changes in Tropical Cyclone Size, Accumulated Cyclone Energy and Power Dissipation Index in a Warmer Climate. *Oceans* **2021**, *2*, 688–699. https://doi.org/10.3390/oceans2040039

Academic Editors: Hiroyuki Murakami and Diego Macías

Received: 8 May 2021
Accepted: 7 September 2021
Published: 11 October 2021

Publisher's Note: MDPI stays neutral with regard to jurisdictional claims in published maps and institutional affiliations.

1. Introduction

With the development of the HighResMIP subproject of the 6th version of the Coupled Model Intercomparison Project, the multi-decadal simulation of global and basin scale tropical cyclone statistics has become mainstream [1]. These models, with horizontal resolutions ranging from 50–20 km can be considered "tropical cyclone permitting" or at least "tropical cyclone-like permitting" as storms are produced in these simulations that bear some similarities to actual tropical cyclones, such as high radial winds, low central pressures and warm central cores [2–7]. Observed patterns and seasonality of cyclogenesis and resulting cyclone tracks can be reasonably reproduced using prescribed sea surface temperatures as a lower boundary condition [5,8] with errors in these statistics manifested by a variety of factors often traceable to subgrid parameterizations. Indeed, as high performance computing platforms edge towards the exascale, some of these model deficiencies, in particular parameterized deep cumulus convection processes, can be ameliorated by yet further increases in horizontal resolution [9].

However, given current computational limitations, HighResMIP-class models are the currently available tool to perform the multi-realization, multi-decadal simulations able to inform about the effect of global warming on tropical cyclone statistics. A recent pair of expert team studies notes that there remains much uncertainty about detectible and attributable changes in observed tropical cyclone statistics [10], even in their projected future changes under significantly more warming than has occurred to date [11]. The first report finds that an observed poleward shift of tropical cyclones in the Northwestern Pacific is "highly unusual compared to expected natural variability" but casts doubt on whether any other observed tropical cyclone properties are detectible, much less attributable to anthropogenic climate change. However, a number of other event attribution studies found that precipitation in individual tropical cyclone has been increased due to warmer sea surface temperatures with low estimates of scaling with temperature increases, according to the Clausius–Clapeyron relationship and best estimates of significantly higher scaling [12–17]. While these studies are not formal detection and attribution studies, confidence that there is a human influence on tropical cyclone precipitation is enhanced by the demonstration

79

that eastern US hurricane precipitation patterns and magnitude can be well represented by models at HighResMIP-class resolutions [18] and in general by previous detection and attribution studies on extreme precipitation [19,20]. Event attribution changes have also investigated anthropogenic changes in tropical cyclone wind speeds [17,21] with very clear future increases but less conclusive findings about the influence at current warming levels [22].

Projections from HighResMIP-class models suggest profound changes in tropical cyclone statistics but with significant uncertainties. The intensity of the most powerful storms, as measured by instantaneous maximum wind speed or minimum central pressure, increases in nearly all of these models with warmer temperatures [5,23–25]. This is very carefully stated in the expert team assessment by Knutson et al. (2019) [11]. As "For TC intensity, 10 of 11 authors had at least medium-to-high confidence that the global average will increase. The mechanism for such a change is straightforward. Intense tropical cyclones occur when ambient wind shear is low and humidity and sea surface temperatures are high. As there will be periods in future warmer climates where wind shear is favorably low, a larger amount of latent and sensible heat energy is available for the storm's kinetic energy. However, this very carefully crafted statement reflects the uncertainty in the number of future tropical storms. Most of the HighResMIP-class models project a decrease in the total number of tropical storms with global warming. But there is substantial variability across models and if the decrease in total number of storms is very large, the number of intense storms may decrease. Hence, another way of stating the expert team assessment is that the fraction of intense tropical cyclones across all tropical storms is expected to increase whether or not the actual number of intense storms increases or decreases. However, the fraction of storms deemed intense is not particularly relevant to impacts, thus motivating this study to examine intensity metrics with nonlinear dependences on peak wind speed.

While the Saffir–Simpson category scale is routinely used to communicate to the public the imminent danger posed by tropical cyclones, more comprehensive alternative scales have been proposed but not widely adopted [26–28]. While interpreted by the impact of selected damage types, the Saffir–Simpson scale is actually defined by 1 min average peak near surface wind exceedances over fixed thresholds. From a detection and attribution point of view, this selection of a pointwise peak from an effectively instantaneous and pointwise quantity may be too noisy to readily ascertain any human influence on tropical cyclone intensity until global warming levels are much larger than present.

In this paper, a selection of other metrics of tropical cyclone intensity are examined with attention paid to their changes as temperatures increase, if any. Projections of future responses to global warming levels larger than that currently realized in the real world can inform us to what changes to expect or at least what to look for in the observations. For these purposes, this study uses simulations from a tropical cyclone permitting model with a strong negative response in global tropical cyclone frequency to warmer global temperatures. These metrics, storm size, accumulated cyclone energy and power dissipation index, are chosen to be more integrative of the entire storm lifecycle than simply counting annual storms in each Saffir–Simpson category. The focus here is only on global quantities but it is recognized that the Northwestern Pacific dominates the global average of most tropical cyclone statistics. Indeed, there is no guarantee that tropical cyclone activity will respond to warming by the same amount or direction across different ocean basins as not only is the warming of the ocean non-uniform, the changes in other tropical cyclogenesis precursors are also non-uniform as are the changes in large scale influences on subsequent tropical cyclone paths and development.

2. The CAM5 Climate Model Setup and Its Tropical Storm Frequency Response on the SAFFIR-Simpson Scale to Warming

The community atmospheric model, version 5.1 (CAM5.1) is a global atmospheric general circulation model with prescribed sea surface temperatures and sea ice concentrations (Neale et al., 2012). For this study, it has been run using a finite difference dynamics scheme on a regular latitude–longitude mesh of approximately one quarter-degree or

~25 km [29]. Its simulated global annual tropical cyclone frequency is remarkably close to observations although substantial cyclogenesis biases are exhibited in the Northern Pacific Basin [8]. Simulated global annual tropical cyclone frequency has previously been shown in this model to decrease relative to that of the current climate when driven by conditions approximating the stabilized 1.5 and 2C targets of the Paris Agreement [24]. Figure 1 extends these simulations to include a preindustrial global temperature level and a stabilized 3C above preindustrial temperature target. The experimental protocols including sea surface temperatures and sea ice concentrations for the preindustrial (here denoted "Natural") and a present day period 1996–2015 (here denoted "Historical") come from the Climate of the 20th Century (C20C+) Detection and Attribution Project (available online: portal.nersc.gov/c20c (accessed on 6 September 2021)), designed for event attribution [30]. Experimental protocols for the 1.5 and 2C stabilized climates come from the the Half A degree additional warming, Prognosis and Projected Impacts (HAPPI) project [31] denoted here as HAPPI1.5 and HAPPI2.0 respectively. The sea surface temperature boundary conditions and greenhouse gas concentrations for the 3C stabilized climate were calculated from the CMIP5 models in the same way as HAPPI but suitably adjusted for the warmer temperatures. As all of the warmer climate simulations are stabilized against future emissions, their aerosol concentrations are set at the preindustrial levels. Only the present day simulations differ in this regard. All the tropical cyclone statistics presented here are calculated from storm tracks obtained from TECA2, the toolkit for extreme climate analysis [32].

The observed frequency of named tropical storms of all Saffir–Simpson intensities from tropical storm to category 5 is about 86 storms per year with an interannual standard deviation of 9.6. From Figure 1, the model under observed boundary conditions produces about 73 storms per year with an interannual standard deviation of 9. Multiple realizations of each temperature scenario were produced. Five simulations of the historical period were concatenated resulting in 100 total simulated years, approximating a stable climate. Ensemble sizes of all the configurations are shown as numbers within the bars of Figure 1. The error bars shown in Figure 1 represent the standard errors calculated using these ensemble sizes. Figure 1 shows that the CAM5.1 model exhibits a strong decrease in storm frequency as the climate warms.

The left panel of Figure 2 reveals that this simulated change in storm frequency varies with Saffir–Simpson categories and the bulk of the decrease in total storm frequency stems from the weaker categories of tropical storm (here denoted as category 0) and category 1. Category 5 storms are more frequent in the future warmer climates than in the preindustrial and current climates despite the overall decrease in cyclogenesis. The same statement is true for category 4 when comparing the future to present climates, but the preindustrial climate actually produced more storms in all other categories than the present day climate. This change in the distribution of peak storm intensities will influence changes in other more integrative intensity metrics.

The right panel of Figure 2 shows the variation in the fraction of annual average storm counts across Saffir-Simpson categories for the various global warming levels. This reveals a somewhat clearer climate change signal, especially for intense tropical cyclones and supports the conservative conclusions of the expert team assessment [11]. It is worth mentioning here that the cleanest comparison is between the natural and the future warmer simulations as they all have the same aerosol forcings. Neglecting the historical simulations then, the fractional increase in intense tropical cyclones (Categories 4 and 5) is monotonic with warming.

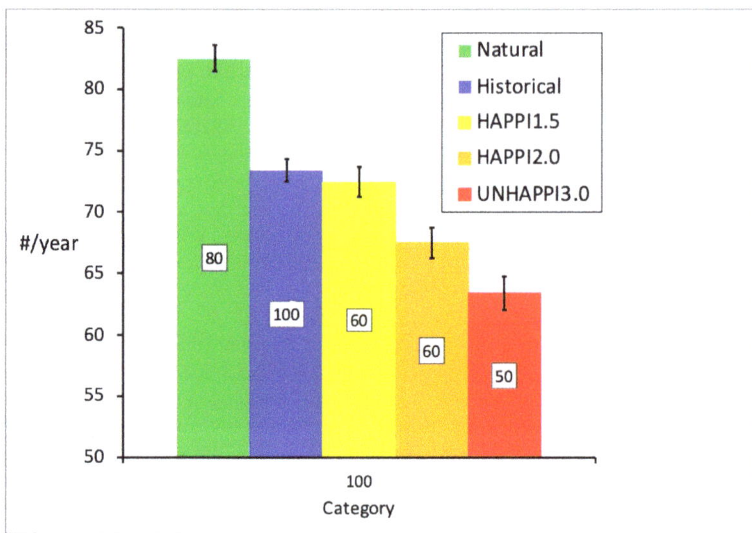

Figure 1. Annual number of all tropical storms (TS-cat5) as simulated by CAM5.1 at various global warming levels. Numbers in the centers are the number of simulated years for each numerical experiment. Error bars indicate standard error.

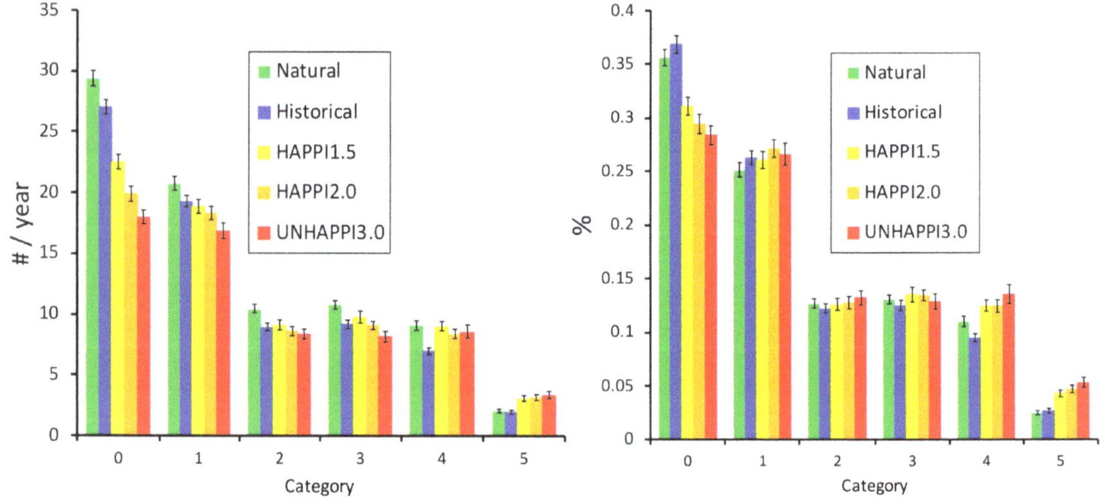

Figure 2. (**Left**) Annual number of tropical storms by category as simulated by CAM5.1 at various global warming levels. (**Right**) Fraction of tropical storms by category as in the left panel. Error bars indicate standard error.

3. Storm Size

Chavas et al., 2015 [33] developed a theoretical model of the radial structure of the low-level tropical cyclone wind field by numerically solving a Riccati equation that relates the radial gradient of the absolute angular momentum and wind speed at a given radius. The spatial distribution of observed storm size from this model using a wind speed of 12 m/s to represent maximum storm extent was shown to agree well with the QuikSCAT Tropical Cyclone Radial Structure database [34]. While this definition of outer storm size radius would provide a good model evaluation metric, here we make a different choice based on much higher wind speeds to define a metric more relevant to impacts. Table 1 shows the Chavas radius as simulated by the CAM5.1 under present day climate conditions for wind speeds defined as the lower bounds of the Saffir–Simpson categories averaged over all the storms in each of the categories. Averages are performed on the three hourly TECA2 output and aggregated over category at that instant. The table is arranged so that for storm of a given category, average wind speed radii are provided at its strongest rating and for all lower categories. Comparison along the diagonal of Table 1 immediately reveals a model of structural weakness. For all tropical cyclone categories greater than 1, the Chavas radius at its rated wind category is only slightly larger than the climate model's horizontal grid resolution (~25 km). Maximum wind radius is then likely constrained to be close to that resolution and hence any simulated tropical cyclone simply revolves around an eye of a single grid cell. Similar tables for the other CAM5.1 simulations are shown in the supplementary materials.

Table 1. Chavas radius (km) for wind speeds thresholds from the CAM5.1 present day (Historical) simulations as a function of instantaneous Saffir–Simpson categorization. Wind speed thresholds are shown in the second row (m/s). The large volume of data renders standard errors miniscule. The radii (r_n) denote the average radius at wind speeds corresponding to the category n threshold. For instance, the average radius (r_1) of category 4 storms at 33m/s is 62.7 km.

	r_0	r_1	r_2	r_3	r_4	r_5
Wind Speed Threshold (m/s)	18	33	43	50	58	70
Category						
TS (cat 0)	48.3	-	-	-	-	-
1	80.9	31.6	-	-	-	-
2	103.1	41.0	27.7	-	-	-
3	119.8	51.4	34.1	28.6	-	-
4	141.1	62.7	44.6	35.6	29.5	-
5	160.4	73.2	52.4	43.4	34.2	26.7

Despite this resolution limitation of the model presented here (and likely HighResMIP-class models in general), it is informative to examine if climate change introduces any change in tropical cyclone average size. The left panel of Figure 3 reveals that there is no consistent change in the average radii of hurricane force winds (category 1 or 33 m/s) while the right panel reveals similarly for the average radii of major hurricane force winds (category 3 or 50 m/s). A similar conclusion is obtained by examining the Chavas radius tables for various global warming levels in the supplementary materials. Hence, at least in this model, climate change does not alter the average wind speed radial distribution of a simulated tropical cyclone of a specified Saffir–Simpson rating. Further evidence that structure of storms at a specified intensity does not change with warming is provided by examining the relationship between maximum surface wind speed and minimum central pressure, which also was found to be similarly unaffected [24]. In plain language for instance, a category 3 storm in a much warmer future world is not larger nor smaller than a category 3 storm in a preindustrial world despite changes in atmospheric structure, such as lapse rates.

This is not to say that the area impacted by hurricane or major hurricane force winds during an entire season does not change with global warming as storm frequencies at different categories will likely change as, for example, in Figure 2. Furthermore, although high resolution event attribution modeling studies have revealed structural changes in individual tropical cyclones as they are warmed [17], such storms are also intensified, which is not inconsistent with the conclusion shown in Figure 3 or the conclusion drawn above. Hence, while average Chavas radius may be a good model performance metric, it is not by itself a good climate change metric.

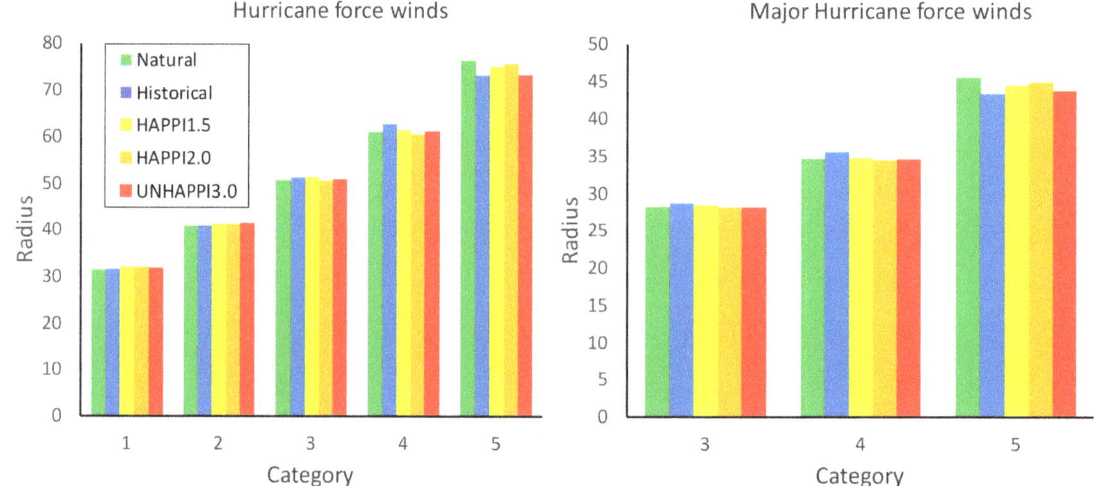

Figure 3. (**Left**) Average radii (km) of hurricane force winds (33 m/s) across Saffir–Simpson categories and global warming levels as simulated by the CAM5.1. (**Right**) Similar but for major hurricane force winds (50 m/s).

4. Accumulated Cyclone Energy Index (ACE)

The accumulated cyclone energy index (ACE) is obtained by summing the square of the peak near surface wind speeds every 6 h over the lifetime of a tropical cyclone. It is commonly used to describe both individual storms as well as seasonal tropical cyclone activity in individual basins or globally. Despite its name, ACE is an index of accumulated pointwise quantities and not a measure of total storm kinetic energy. It is another useful metric, together with storm count, to describe the variations in seasonal tropical cyclone activity. Basin wide ACE statistics have been used as a model validation metric [35] revealing that the CAM5.1 simulated distribution of ACE in North Pacific is skewed toward excess in the eastern part of the basin similar to storm counts [8]. Globally, the present day CAM5.1 simulation is about 20% higher than the observed average over 1995–2015 of 750 ACE units (10^4 knots^2). The left panel of Figure 4 shows global ACE from the CAM5.1 simulations, revealing that present day simulated ACE is both less than in the cooler preindustrial climate and in the warmer future climates amidst substantial uncertainty from interannual variability. This CAM5.1 projection is consistent with similarly inconclusive total global ACE projections [36].

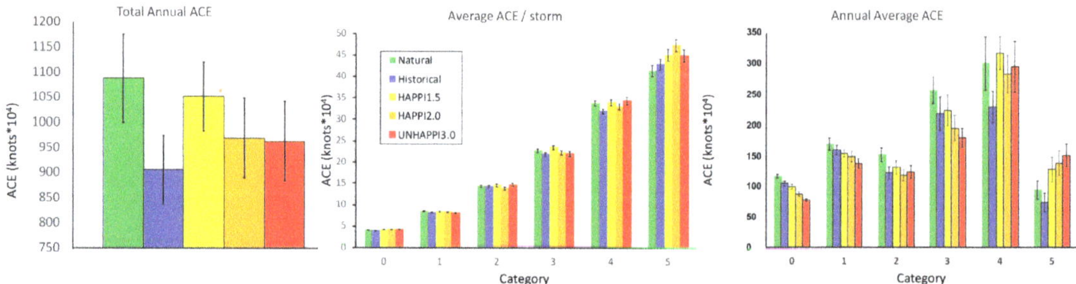

Figure 4. (**Left**) Average annual global accumulated cyclone energy index (ACE) as simulated by CAM5.1 at various global warming levels. Error bars indicate standard error. (**Center**) Average ACE per storm across Saffir–Simpson categories as simulated by CAM5.1 at various global warming levels. (**Right**) Average annual global ACE across Saffir–Simpson categories as simulated in the center panel. Error bars indicate standard error.

The reason for this non-uniform change in average global ACE from one warming level to another is a result of the convolution of changes in tropical cyclone frequency and their distribution across wind speed intensities (Figures 1 and 2). The center panel of Figure 4 shows the average ACE for storms according to their assigned peak category. From tropical storm intensity to category 4, average storm ACE does not change with global warming amount. Similar to the conclusion about this model's storm size from Figure 3 and related tables, climate change does not change the average ACE of storms within these categories. This might be a bit surprising as this model was shown to exhibit longer tropical storm lifetimes with an associated increased poleward track density as the climate warms [24]. However, ACE depends on the square of the instantaneous peak wind speed and the bulk of a storm's ACE is accumulated during its time spent in or near its strongest rated category. This may suggest that the intensification and subsequent decay of tropical cyclones may be unaffected by global warming until late in their lifetimes. However, confidence in this level of detail drawn from a HighResMIP-class model should be very low as this aspect of tropical cyclone development is notoriously difficult to simulate [37]. Average storm ACE within category 5 generally increases with warming as that category is open ended and reflects that the most intense storms become yet more intense with warmer sea surface temperatures. While the UNHAPPI3.0 simulation appears to be an exception, this is also the case with the fewest number of simulated years and the highest uncertainty.

The distribution of simulated annual global average ACE across the Saffir–Simpson categories is shown in the right panel of Figure 4 and reveals that most of the total ACE comes from category 4 storms with the second largest contribution coming from category 3 storms at any global warming level. Changes with warming level in this figure are largely controlled by changes in the tropical cyclone categories counts of Figure 2.

5. Power Dissipation Index (PDI)

The power dissipation index (PDI) is similar to ACE in that it is a (partial) measure of total storm intensity. Instead of accumulating the square of the peak surface wind speed, PDI accumulates the cube of the peak wind speed. Analogous to friction energy applied by flow to a surface, the cube of peak surface wind speed is more closely related to economic damages than tropical cyclone frequency itself [38]. Moreover, similar to ACE, seasonal accumulated PDI will be more influenced by the most intense storms only more so due to the higher nonlinear dependence on peak surface wind speed. Similar to Figure 4, Figure 5 shows annual average global PDI (left), average PDI per storm (center) and average annual global PDI (right) as simulated by the CAM5.1 at various global warming levels. Similar to ACE, the largest contributor to total simulated PDI comes from category 4 storms. However, in this case for the warmer climate conditions, category 5 storms can contribute as much or more to total PDI as category 3 storms reflecting both the intensification of the largest storms and PDI's cubic dependence on peak wind speed.

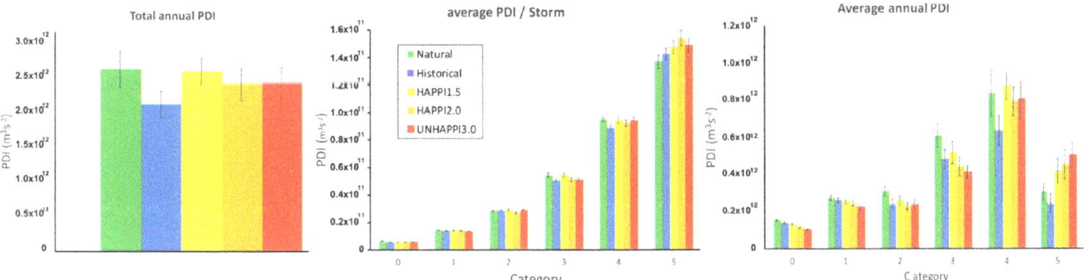

Figure 5. (**Left**) Average annual global power dissipation index (PDI) as simulated by CAM5.1 at various global warming levels. Error bars indicate standard error. (**Center**): Average PDI per storm across Saffir-Simpson categories as simulated by CAM5.1 at various global warming levels. (**Right**) Average annual global PDI across Saffir–Simpson categories as simulated in the center panel. Error bars indicate standard error.

6. Discussion

This study explores global average storm size, accumulated cyclone energy (ACE) and power dissipation index (PDI) as alternatives to simple counting by Saffir-Simpson scale for the detection, attribution and projection of changes in tropical cyclone activity as the planet warms due to anthropogenic influences. As observations are limited, a high resolution (~25 km) global atmospheric general circulation model is used as a tool to examine what changes, if any, might be robust and possibly contained in the actual climate system. While convection permitting models (~4 km or finer) would be a preferable tool for analyzing changes in storm structural statistics, computational constraints preclude the formation of ensemble multi-decadal simulations necessary to extract climate change signals, if any, from the underlying noise.

Simulated changes in the total global annual average ACE and PDI are not found to be robust to global warming. This is largely a result of offsetting changes in overall decreasing storm counts but increasing average intensity. However, it is entirely possible that regional changes in these metrics may be robust if regional changes in storm frequency are substantially different than what they are globally. Indeed, substantial ACE and PDI increases in the North Atlantic have been observed [38,39] and the model exhibits North Atlantic ACE increases but not in the North Pacific [24]. When sorted on the Saffir–Simpson scale, only the highest unbounded category exhibits increased ACE and PDI with warming for the average storm within a category despite an increase in simulated storm duration across categories.

Average instantaneous storm size, as measured by the Chavas radius at surface wind speeds at hurricane (33 m/s) and major hurricane force (50 m/s), is also found not to change with global warming (Figure 3) as might be expected, although model resolution is not high enough to adequately capture eye wall details. Previous studies have focused on average outer storm size, typically measured at 12 m/s or slower [33,40]. Indeed, outer storm size may be a more appropriate detection and attribution metric than the inner radii of Figure 3 as it may be more readily observable and less affected by eyewall processes, although it is less relevant for wind damage impacts. However, climate change projections of average outer storm size are conflicting as Yamada et al. (2017) [41] found an increase with global warming in a 14 km model but Knutson et al. (2015) [42] found no change in the media outer storm size used downscaled CMIP5 models with a 6 km hurricane forecast model. Comparison of r_0 in the first column of Table 1 to that in the tables in the supplementary materials also suggests no change in storm radii at 18 m/s.

There is evidence from event attribution studies that the radial structure of tropical storm precipitation may be affected by global warming [12,17]. However, Figure 3 and the tables show that average radii for each wind speed considered within the Saffir–Simpson categories is not sensitive to global warming. Consistent with previous analyses showing no change with warming in the peak wind speed to pressure minima relationship [24,43,44], this null result suggests that because the wind speed category bounds are relatively narrow, storms for any specified peak wind speed are structurally similar regardless of global warming level. The structural changes seen in event attribution studies are then simply the result in shifts in wind speeds rather than some fundamental structural change. One then might expect a change in the wind speed radii for the unbounded category 5, as only for that category does the average wind speed change with warming. However, the present model's limitations in simulating eyewall processes is especially important when considering such intense storms and the present null result in this case should not be considered definitive.

Dividing global annual average ACE and PDI according to Saffir–Simpson categories yields results very similar to simple counting by categories with general decreases at the lower categories and increases in the highest one. This suggests then that a more robust climate change metric might be exceedances over a high threshold rather than averages. Figure 6 shows the exceedances of each climate scenario over a threshold selected from the historical simulation and reveals little difference between ACE (middle) and PDI (right). The left panel of Figure 6 shows exceedances of storm peak wind speed. However, this panel is not directly comparable to the other two as is calculated from the storm maximum wind speed over entire storm lifetime and as such is far more extreme than the storm integrated quantities of ACE and PDI. It does not appear that accumulating tropical cyclone intensity metrics over the course of storms and seasons adds significant clarity to disentangling the issue of decreasing storm frequency but increases in the tail of the wind speed distribution.

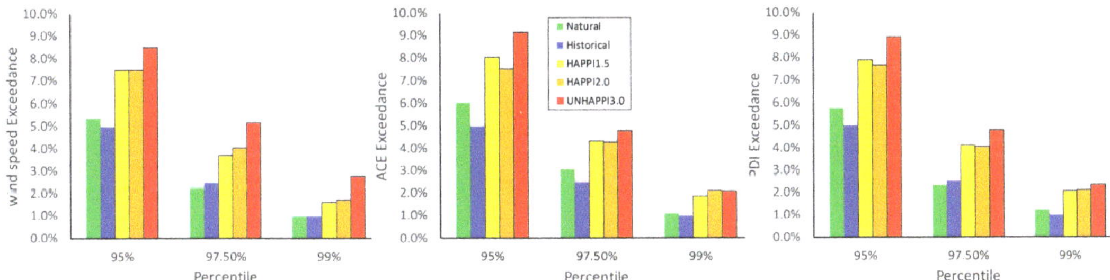

Figure 6. Exceedance over 95, 97.5 and 99th thresholds selected from the Historical CAM5.1 simulation. (**Left**) Maximum storm peak wind. (**Middle**) Storm total ACE. (**Right**) Storm total PDI.

This attempt to find a better metric than global tropical cyclone frequency for climate change detection, attribution and projection produces mixed results. The 6th Assessment Report of the Intergovernmental Panel on Climate Change [45] concluded with high confidence that the fraction of tropical cyclones that achieve category 4 wind speeds or higher would increase with further global warming but made no statement about the number of such intense tropical cyclones. Consistent with Knutson et al. (2019), the assessment recognized that available model projections, including the HighResMIP models, vary greatly in projected decreased global total tropical storm frequency with warming, if any. Hence, if the actual decrease in total tropical storm frequency were to be small and intensification large enough, there would be more intense tropical storms. However, if the decrease in total frequency is large enough, there would be fewer intense tropical storms. Indeed, the trend in intense storm frequency might not even be monotonic with increases at low levels of global warming but decreases at higher levels due to this contention between increased intensification and decreased cyclogenesis. In the context of the current study, this structural uncertainty in future projections of the distribution of tropical storm intensity carries over to future projections of both global ACE and PDI.

While storm size, ACE and PDI are important climate model performance evaluation metrics [16], this combined effect of global warming of decreasing storm count but increasing the intensities of the strongest storms complicates constructing a robust global metric that might exhibit a change given enough data to reduce internal variability. While this study used a climate model that produced between 50 and 100 years of tropical cyclones under stabilized climate scenarios, the real world is a more complex transient system with smaller available data set sizes. Present day exceedance of a contemporary 95th percentile global intensity threshold would result in about four storms annually. Due to the large natural variability of peak tropical storm intensities, confident detection and attribution of the effect of global warming on tropical cyclone intensity statistics relevant to impacts may not be realized with the simple global statistics considered here until far into the future. However, regional versions of these metrics or other even more complex metrics, such as the distribution of storm tracks, storm duration and translational speed, may be more promising.

Supplementary Materials: The following are available online at https://www.mdpi.com/article/10.3390/oceans2040039/s1: Table S1. Chavas radius (km) for wind speeds thresholds from the CAM5.1 preindustrial simulation (NATURAL) as a function of instantaneous Saffir–Simpson categorization; Table S2. Chavas radius (km) for wind speeds thresholds from the CAM5.1 1.5 °C above preindustrial simulation (HAPPI1.5) as a function of instantaneous Saffir–Simpson categorization; Table S3. Chavas radius (km) for wind speeds thresholds from the CAM5.1 2 °C above preindustrial simulation (HAPPI2.0) as a function of instantaneous Saffir–Simpson categorization; Table S4. Chavas radius (km) for wind speeds thresholds from the CAM5.1 3 °C above preindustrial simulation (UNHAPPI3.0) as a function of instantaneous Saffir–Simpson categorization.

Funding: This research was supported by the Director, Office of Science, Office of Biological and Environmental Research of the U.S. Department of Energy under Contract No. DE340AC02-05CH11231 and funding from the Regional and Global Model Analysis program.

Institutional Review Board Statement: Not applicable.

Informed Consent Statement: Not applicable.

Data Availability Statement: All high resolution CAM5.1 data used in this paper is available via the Climate of the 20th Century data portal, portal.nersc.gov/c20c (accessed on 6 September 2021).

Acknowledgments: The author thanks Burlen Loring (LBNL) for incorporating these metrics into TECA2 and Paul Ullrich (UC Davis) and Kevin Reed (Stony Brook) for helpful suggestions. This research was supported by the Director, Office of Science, Office of Biological and Environmental Research of the U.S. Department of Energy under Contract No. DE340AC02-05CH11231 and funding from the Regional and Global Model Analysis program. This document was prepared as an account of work sponsored by the United States Government. While this document is believed to contain correct information, neither the United States Government nor any agency thereof, nor the Regents of the University of California, nor any of their employees, makes any warranty, express or implied, or assumes any legal responsibility for the accuracy, completeness, or usefulness of any information, apparatus, product, or process disclosed, or represents that its use would not infringe privately owned rights. Reference herein to any specific commercial product, process, or service by its trade name, trademark, manufacturer, or otherwise, does not necessarily constitute or imply its endorsement, recommendation, or favoring by the United States Government or any agency thereof, or the Regents of the University of California. The views and opinions of authors expressed herein do not necessarily state or reflect those of the United States Government or any agency thereof or the Regents of the University of California.

Conflicts of Interest: The author declares no conflict of interest.

References

1. Haarsma, R.J.; Roberts, M.J.; Vidale, P.L.; Senior, C.A.; Bellucci, A.; Bao, Q.; Chang, P.; Corti, S.; Fučkar, N.S.; Guemas, V.; et al. High resolution model intercomparison project (HighResMIP v1.0) for CMIP6. geosci. *Model Dev.* **2016**, *9*, 4185–4208. [CrossRef]
2. Wehner, M.F.; Bala, G.; Duffy, P.; Mirin, A.A.; Romano, R. Towards direct simulation of future tropical cyclone statistics in a high-resolution global atmospheric model. *Adv. Meteorol.* **2010**, *2010*, 915303. [CrossRef]
3. Reed, K.A.; Bacmeister, J.T.; Rosenbloom, N.A.; Wehner, M.F.; Bates, S.C.; Lauritzen, P.H.; Truesdale, J.E.; Hannay, C. Impact of the dynamical core on the direct simulation of tropical cyclones in a high-resolution global model. *Geophys. Res. Lett.* **2015**, *42*, 3603–3608. [CrossRef]
4. Strachan, J.; Vidale, P.L.; Hodges, K.; Roberts, M.; Demory, M.-E. Investigating global tropical cyclone activity with a hierarchy of AGCMs: The role of model resolution. *J. Clim.* **2013**, *26*, 133–152. [CrossRef]
5. Roberts, M.J.; Camp, J.; Seddon, J.; Vidale, P.L.; Hodges, K.; Vanniere, B.; Mecking, J.; Haarsma, R.; Bellucci, A.; Scoccimarro, E.; et al. Impact of model resolution on tropical cyclone simulation using the HighResMIP–PRIMAVERA multimodel ensemble. *J. Clim.* **2020**, *33*, 2557–2583. [CrossRef]
6. Murakami, H.; Vecchi, G.A.; Underwood, S.; Delworth, T.L.; Wittenberg, A.T.; Anderson, W.G.; Chen, J.H.; Gudgel, R.G.; Harris, L.M.; Lin, S.J.; et al. Simulation and prediction of category 4 and 5 hurricanes in the high-resolution GFDL HiFLOR coupled climate model. *J. Clim.* **2015**, *28*, 9058–9079. [CrossRef]
7. Reed, K.A.; Jablonowski, C. Impact of physical parameterizations on idealized tropical cyclones in the Community Atmosphere Model. *Geophys. Res. Lett.* **2011**, *38*. [CrossRef]
8. Wehner, M.F.; Reed, K.A.; Li, F.; Bacmeister, J.; Chen, C.-T.; Paciorek, C.; Gleckler, P.; Sperber, K.; Collins, W.D.; Andrew, G.; et al. The effect of horizontal resolution on simulation quality in the Community Atmospheric Model, CAM5.1. *J. Adv. Model. Earth Syst.* **2014**, *6*, 980–997. [CrossRef]

9. Satoh, M.; Tomita, H.; Yashiro, H.; Kajikawa, Y.; Miyamoto, Y.; Yamaura, T.; Miyakawa, T.; Nakano, M.; Kodama, C.; Noda, A.T.; et al. Outcomes and challenges of global high-resolution non-hydrostatic atmospheric simulations using the K computer. *Prog. Earth Planet. Sci.* **2017**, *4*, 13. [CrossRef]
10. Knutson, T.; Camargo, S.J.; Chan, J.C.L.; Emanuel, K.; Ho, C.-H.; Kossin, J.; Mohapatra, M.; Satoh, M.; Sugi, M.; Walsh, K.; et al. Tropical cyclones and climate change assessment: Part I: Detection and attribution. *Bull. Am. Meteorol. Soc.* **2020**, *100*, 1987–2007. [CrossRef]
11. Knutson, T.; Camargo, S.J.; Chan, J.C.L.; Emanuel, K.; Ho, C.-H.; Kossin, J.; Mohapatra, M.; Satoh, M.; Sugi, M.; Walsh, K.; et al. Tropical Cyclones and Climate Change Assessment: Part II. Projected Response to Anthropogenic Warming. *Bull. Am. Meteorol. Soc.* **2019**, *101*, E303–E322. [CrossRef]
12. Reed, K.A.; Stansfield, A.M.; Wehner, M.F.; Zarzycki, C.M. Forecasted attribution of the human influence on Hurricane Florence. *Sci. Adv.* **2020**, *6*. [CrossRef] [PubMed]
13. van Oldenborgh, G.J.; van der Wiel, K.; Sebastian, A.; Singh, R.; Arrighi, J.; Otto, F.; Haustein, K.; Li, S.; Vecchi, G.; Cullen, H. Attribution of extreme rainfall from Hurricane Harvey, August 2017. *Environ. Res. Lett.* **2017**, *12*, 124009. [CrossRef]
14. Wang, S.Y.S.; Zhao, L.; Yoon, J.H.; Klotzbach, P.; Gillies, R.R. Quantitative attribution of climate effects on Hurricane Harvey's extreme rainfall in Texas. *Environ. Res. Lett.* **2018**. [CrossRef]
15. Risser, M.D.; Wehner, M.F. Attributable human-induced changes in the likelihood and magnitude of the observed extreme precipitation during hurricane harvey. *Geophys. Res. Lett.* **2017**, *44*, 12457–12464. [CrossRef]
16. Zarzycki, C.M.; Ullrich, P.A.; Reed, K.A. Metrics for Evaluating Tropical Cyclones in Climate Data. *J. Appl. Meteorol. Climatol.* **2021**, *60*, 643–660. [CrossRef]
17. Patricola, C.M.; Wehner, M.F. Anthropogenic influences on major tropical cyclone events. *Nature* **2018**, *563*, 339–346. [CrossRef]
18. Stansfield, A.M.; Reed, K.A.; Zarzycki, C.M.; Ullrich, P.A.; Chavas, D.R. Assessing Tropical Cyclones' Contribution to Precipitation over the Eastern United States and Sensitivity to the Variable-Resolution Domain Extent. *J. Hydrometeorol.* **2020**, *21*, 1425–1445. [CrossRef]
19. Min, S.-K.; Zhang, X.; Zwiers, F.W.; Hegerl, G.C. Human contribution to more-intense precipitation extremes. *Nature* **2011**, *470*, 378. [CrossRef]
20. Zhang, X.; Wan, H.; Zwiers, F.W.; Hegerl, G.C.; Min, S.-K. Attributing intensification of precipitation extremes to human influence. *Geophys. Res. Lett.* **2013**, *40*, 5252–5257. [CrossRef]
21. Takayabu, I.; Hibino, K.; Sasaki, H.; Shiogama, H.; Mori, N.; Shibutani, Y.; Takemi, T. Climate change effects on the worst-case storm surge: A case study of Typhoon Haiyan. *Environ. Res. Lett.* **2015**, *10*, 064011. [CrossRef]
22. Wehner, M.F.; Reed, K.A.; Zarzycki, C.M. High-resolution multi-decadal simulation of tropical cyclones, Hurricanes and Climate Change. In *Hurricanes and Climate Change*; Springer: Cham, Switzerland, 2017. [CrossRef]
23. Oouchi, K.; Yoshimura, J.; Yoshimura, H.; Mizuta, R.; Kusonoki, S.; Noda, A. Tropical cyclone climatology in a global-warming climate as simulated in a 20 km-mesh global atmospheric model: Frequency and wind intensity analyses. *J. Meteorol. Soc. Japan. Ser. II* **2006**, *84*, 259–276. [CrossRef]
24. Wehner, M.F.; Reed, K.A.; Loring, B.; Stone, D.; Krishnan, H. Changes in tropical cyclones under stabilized 1.5 and 2.0 °C global warming scenarios as simulated by the Community Atmospheric Model under the HAPPI protocols. *Earth Syst. Dyn.* **2018**, *9*, 187–195. [CrossRef]
25. Bhatia, K.; Vecchi, G.; Murakami, H.; Underwood, S.; Kossin, J. Projected response of tropical cyclone intensity and intensification in a global climate model. *J. Clim.* **2018**, *31*, 8281–8303. [CrossRef]
26. Bloemendaal, N.; de Moel, H.; Mol, J.M.; Bosma, P.R.M.; Polen, A.N.; Collins, J.M. Adequately reflecting the severity of tropical cyclones using the new Tropical Cyclone Severity Scale. *Environ. Res. Lett.* **2021**, *16*, 14048. [CrossRef]
27. Bosma, C.D.; Wright, D.B.; Nguyen, P.; Kossin, J.P.; Herndon, D.C.; Shepherd, J.M. An Intuitive Metric to Quantify and Communicate Tropical Cyclone Rainfall Hazard. *Bull. Am. Meteorol. Soc.* **2020**, *101*, E206–E220. [CrossRef]
28. Song, J.Y.; Alipour, A.; Moftakhari, H.R.; Moradkhani, H. Toward a more effective hurricane hazard communication. *Environ. Res. Lett.* **2020**, *15*, 64012. [CrossRef]
29. Bacmeister, J.T.; Wehner, M.F.; Neale, R.B.; Gettelman, A.; Hannay, C.; Lauritzen, P.H.; Caron, J.M.; Truesdale, J.E. Exploratory high-resolution climate simulations using the community atmosphere model (CAM). *J. Clim.* **2014**, *27*, 3073–3099. [CrossRef]
30. Stone, D.A.; Christidis, N.; Folland, C.; Perkins-Kirkpatrick, S.; Perlwitz, J.; Shiogama, H.; Wehner, M.F.; Wolski, P.; Cholia, S.; Krishnan, H.; et al. Experiment design of the international CLIVAR C20C+ detection and attribution project. *Weather Clim. Extrem.* **2019**, *24*, 100206. [CrossRef]
31. Mitchell, D.; AchutaRao, K.; Allen, M.; Bethke, I.; Beyerle, U.; Ciavarella, A.; Forster, P.M.; Fuglestvedt, J.; Gillett, N.; Haustein, K.; et al. Half a degree additional warming, prognosis and projected impacts (HAPPI): Background and experimental design. *Geosci. Model Dev.* **2017**, *10*, 571–583. [CrossRef]
32. Prabhat; Rübel, O.; Byna, S.; Wu, K.; Li, F.; Wehner, M.; Bethel, W. TECA: A parallel toolkit for extreme climate analysis. In *Procedia Computer Science*; Elsevier: Amsterdam, The Netherlands, 2012. [CrossRef]
33. Chavas, D.R.; Lin, N.; Emanuel, K. A model for the complete radial structure of the tropical cyclone wind field. Part I: Comparison with observed structure. *J. Atmos. Sci.* **2015**, *72*, 3647–3662. [CrossRef]
34. Chavas, D.R.; Lin, N.; Dong, W.; Lin, Y. Observed tropical cyclone size revisited. *J. Clim.* **2016**, *29*, 2923–2939. [CrossRef]

35. Shaevitz, D.A.; Camargo, S.J.; Sobel, A.H.; Jonas, J.A.; Kim, D.; Kumar, A.; Larow, T.E.; Lim, Y.-K.; Murakami, H.; Reed, K.A.; et al. Characteristics of tropical cyclones in high-resolution models in the present climate. *J. Adv. Model. Earth Syst.* **2014**, *6*. [CrossRef]
36. Guimarães, S.O. Climate models accumulated cyclone energy analysis. In *Current Topics in Tropical Cyclone Research*; Lupo, A., Ed.; IntechOpen: London, UK, 2020. [CrossRef]
37. Judt, F.; Chen, S.S. Predictability and dynamics of tropical cyclone rapid intensification deduced from high-resolution stochastic ensembles. *Mon. Weather Rev.* **2016**, *144*, 4395–4420. [CrossRef]
38. Emanuel, K. Increasing destructiveness of tropical cyclones over the past 30 years. *Nature* **2005**, *436*, 686–688. [CrossRef]
39. Murakami, H.; Li, T.; Hsu, P.C. Contributing factors to the recent high level of accumulated cyclone energy (ACE) and power dissipation index (PDI) in the North Atlantic. *J. Clim.* **2014**, *27*, 3023–3034. [CrossRef]
40. Chavas, D.R.; Reed, K.A. Dynamical aquaplanet experiments with uniform thermal forcing: System dynamics and implications for tropical cyclone genesis and size. *J. Atmos. Sci.* **2019**, *76*, 2257–2274. [CrossRef]
41. Yamada, Y.; Satoh, M.; Sugi, M.; Kodama, C.; Noda, A.T.; Nakano, M.; Nasuno, T. Response of tropical cyclone activity and structure to global warming in a high-resolution global nonhydrostatic model. *J. Clim.* **2017**, *30*, 9703–9724. [CrossRef]
42. Knutson, T.R.; Sirutis, J.J.; Zhao, M.; Tuleya, R.E.; Bender, M.; Vecchi, G.A.; Villarini, G.; Chavas, D. Global Projections of Intense Tropical Cyclone Activity for the Late Twenty-First Century from Dynamical Downscaling of CMIP5/RCP4.5 Scenarios. *J. Clim.* **2015**, *28*, 7203–7224. [CrossRef]
43. Chavas, D.R.; Reed, K.A.; Knaff, J.A. Physical understanding of the tropical cyclone wind-pressure relationship. *Nat. Commun.* **2017**, *8*, 1360. [CrossRef]
44. Satoh, M.; Yamada, Y.; Sugi, M.; Komada, C.; Noda, A.T. Constraint on future change in global frequency of tropical cyclones due to global warming. *J. Meteorol. Soc. Jpn. Ser. II* **2015**, *93*, 489–500. [CrossRef]
45. Seneviratne, S.I.; Zhang, X.; Adnan, M.; Badi, W.; Dereczynski, C.; di Luca, A.; Ghosh, S.; Iskandar, I.; Kossin, J.; Lewis, S.; et al. Weather and Climate Extreme Events in a Changing Climate. In *Climate Change 2021: The Physical Science Basis. Contribution of Working Group I to the Sixth Assessment Report of the Intergovern-Mental Panel on Climate Change*; Masson-Delmotte, V., Zhai, P., Pirani, A., Connors, S.L., Péan, C., Berger, S., Caud, N., Chen, Y., Goldfarb, L., Gomis, M.I., et al., Eds.; Cambridge University Press: Cambridge, UK, 2021, in press.

Article

Response of Tropical Cyclone Frequency to Sea Surface Temperatures Using Aqua-Planet Simulations

Pavan Harika Raavi [1,2,†] and **Kevin J. E. Walsh** [1,2,*]

[1] School of Earth Sciences, University of Melbourne, Parkville, VIC 3010, Australia; praavi@student.unimelb.edu.au

[2] ARC Centre of Excellence for Climate Extremes, University of New South Wales, Sydney, NSW 2052, Australia

[*] Correspondence: kevin.walsh@unimelb.edu.au

[†] Current address: Center for Climate Physics, Institute for Basic Science (IBS), Pusan National University, Busan 46241, Korea.

Abstract: The present study investigates the effect of increasing sea surface temperatures (SSTs) on tropical cyclone (TC) frequency using the high-resolution Australian Community Climate and Earth-System Simulator (ACCESS) model. We examine environmental conditions leading to changes in TC frequency in aqua-planet global climate model simulations with globally uniform sea surface temperatures (SSTs). Two different TC tracking schemes are used. The Commonwealth Scientific and Industrial Research Organization (CSIRO) scheme (a resolution-dependent scheme) detects TCs that resemble observed storms, while the Okubo–Weiss zeta parameter (OWZP) tracking scheme (a resolution-independent scheme) detects the locations within "marsupial pouches" that are favorable for TC formation. Both schemes indicate a decrease in the global mean TC frequency with increased saturation deficit and static stability of the atmosphere. The OWZP scheme shows a poleward shift in the genesis locations with rising temperatures, due to lower vertical wind shear. We also observe an overall decrease in the formation of tropical depressions (TDs) with increased temperatures, both for those that develop into TCs and non-developing cases. The environmental variations at the time of TD genesis between the developing and the non-developing tropical depressions identify the Okubo–Weiss (OW) parameter and omega (vertical mass flux) as significant influencing variables. Initial vortices with lower vorticity or with weaker upward mass flux do not develop into TCs due to environments with higher saturation deficit and stronger static stability of the atmosphere. The latitudinal variations in the large-scale environmental conditions account for the latitudinal differences in the TC frequency in the OWZP scheme.

Keywords: aqua-planet global climate model simulations; Okubo–Weiss zeta parameter; tropical depression; tropical cyclone

Citation: Raavi, P.H.; Walsh, K.J.E. Response of Tropical Cyclone Frequency to Sea Surface Temperatures Using Aqua-Planet Simulations. *Oceans* **2021**, *2*, 785–810. https://doi.org/10.3390/oceans2040045

Academic Editor: Hiroyuki Murakami

Received: 14 April 2021
Accepted: 10 November 2021
Published: 1 December 2021

Publisher's Note: MDPI stays neutral with regard to jurisdictional claims in published maps and institutional affiliations.

1. Introduction

Tropical cyclones (TCs) cause severe socio-economic losses, mainly in coastal communities. TCs are largely controlled by the surrounding environmental conditions and the large-scale dynamical features of the tropics. Both observational and modeling studies have been used to understand the relevant environmental influences and formation mechanisms. A recent generation of climate models has produced simulations with a reasonable geographical distribution and seasonal variability of TCs but have uncertainties in the simulated intensity due to its sensitivity to the representation of sub-grid scale processes and horizontal resolution [1–11]. From these studies, it can be noted that there are differences in the processes governing TC formation versus intensification. For climate change predictions, there is an agreement between the theory and modeling of TC intensity, both predicting an increase with increasing temperatures ([12,13]; see [14], 2020 for a recent review). In contrast, although most modeling studies show a decrease in TC frequency,

91

the precise mechanism leading to these changes is not known. Statistically derived TC genesis potential indices, based on the best statistical fit of TC genesis and climate variables in the current climate, perform well in the current climate, but they do not consistently show a decrease in TC frequency with increasing temperatures [15]. Thus, these indices are not necessarily a representation of changes in TC formation in the future climate. Additionally, there is no generally accepted climate theory for TC formation that quantifies their frequency based on the background climate conditions (e.g., [16]), despite recent progress on this issue [17,18]. These uncertainties in TC frequency projections require us to better understand the quantitative values of prominent environmental variables leading to changes in the TC frequency.

One method for addressing this issue is to use idealized numerical model experiments that remove some of the confounding influences of different variables. Many idealized studies on the environmental conditions leading to the changes in TC characteristics such as formation, size, and intensity due to changing climate conditions have been carried out using high-resolution cloud-resolving models (at a horizontal resolution of ~1 km), with a simpler TC-permitting framework than would generally be included in a typical climate model. For example, the radiative–convective equilibrium (RCE) model framework has been employed in such studies, and those models use idealized boundary conditions that have a range of uniform sea surface temperatures (SSTs), either in an f-plane or non-rotating geometry, with domains of various sizes [19–26]. These studies highlight the use of simplified boundary conditions to understand the role of the planetary vorticity, surface fluxes, radiative feedbacks, and mean thermodynamic environmental controls on different TC characteristics. Studies using large-domain RCE experiments indicate that TC size increases with increasing temperature, while with an increase in Coriolis parameter f, there is a decrease in size and an increase in the intensity. Some studies note that TC size scales with the ratio of the potential intensity [13] to the Coriolis parameter and with the radius of deformation [20,27]. Using idealized RCE experiments, Chavas and Emanuel (2014) [28] altered radiative cooling to determine the TC size variations with scaled potential intensity and radius of deformation, showing more consistent results with the scaled potential intensity than with radius of deformation. Chavas and Emanuel (2014) [28] used specified SSTs in their simulations, and this simplification of the bottom boundary conditions has been implemented in general circulation models (GCMs) to study the changes in TC characteristics and the circulation changes induced by the numerous vortices generated in such idealized simulations. These types of simulations are known as aqua-planet simulations [29,30]. Often, such experiments have no zonal SST asymmetries, with the surface covered with water, employing either a slab ocean or constant specified SSTs.

Merlis et al. (2013) [31] use an aqua-planet GCM simulation with a slab ocean, noting that TC frequency increases with increasing radiative forcing with no change in the cross-equatorial ocean heat flux, due to a shift in the Inter-Tropical Convergence Zone (ITCZ) in the poleward direction. Ballinger et al. (2015) [32] use a GCM with zonally symmetric and inter-hemispherically asymmetric surface forcing with fixed SSTs and slab ocean boundary conditions. They note that the TC frequency response is sensitive to the meridional SST gradients and is directly proportional to the distance of the ITCZ from the equator. These two studies focus on the ITCZ position and its relationship to the genesis frequency, associated with the changes in the atmospheric circulation driven by the underlying asymmetric thermal forcing. Using constant SSTs as lower boundary conditions, Merlis et al. (2016) [33] compare the TC frequency response to SST in GCM simulations both with spherical geometry that includes variation of the Coriolis parameter and with f-plane RCE simulations. They note an inhomogeneous distribution of TCs with an increasing trend from low to high latitudes, with the subtropics as the preferred region. Moreover, in both RCE and GCM simulations, there is a decrease in TC frequency and a poleward shift in the genesis with increasing temperatures. A study by Shi and Bretherton (2014) [34] using a GCM with spherical geometry, constant SST and no insolation shows the formation of multiple TC-like vortices poleward of 10 degrees latitude, with durations longer than two months, considerably longer than those observed in the current terrestrial climate. In addition to the above studies, Reed and Chavas (2015) [35] use aqua-planet experiments with spatially uniform forcing to show that TCs fill the global domain, so that TC numbers and storm size become closely related.

Recently, Chavas and Reed (2019) [17] provide a new theory which bridges the gap between the f-plane RCE simulations and the rotating sphere GCM simulations, and then tests it using uniform aqua-planet experiments with varying planetary rotation rate and size. They note that on the f-plane there is only one relevant length scale for TC formation, which scales with $1/f$. On the sphere, there is a second relevant length scale associated with $\partial f / \partial y$ called the Rhines scale. At high latitudes, the Rhines scale becomes very large, so only the $1/f$ scale remains relevant, and hence this region is dynamically equivalent to an f-plane, supporting many TCs. A recent review by Merlis and Held (2019) [30] on the various aqua-planet studies of TCs, from single TC-permitting models to GCMs, highlights the importance of the idealized model configurations in understanding the mechanisms that control TC activity due to changes in climate conditions. More recently, Walsh et al. (2020) [10] employ different aqua-planet simulations with both constant and meridionally varying SSTs to study the quantitative influence of large-scale climate on the TC frequency changes. They note that the vertical static stability parameter has a significant influence on the TC frequency and acts as a likely important parameter for any climate theory of TC formation.

A recent TC formation theory, the "marsupial pouch" theory, states that a semi-enclosed recirculating region exists within the trough region of tropical waves that protects the vortex from external disruptive influences [36]. Wang et al. (2012) [37] hypothesize that a deep pouch spanning from the lower to the middle troposphere could be sufficient for TC formation. The above studies using idealized model simulations motivate us to analyze the changes in the frequency of the TC-favorable marsupial pouch formation environments and the associated large-scale environmental influences on their formation. The Okubo–Weiss Zeta Parameter (OWZP) TC detection scheme [8,38] detects the locations within marsupial pouches favorable for TC formation. This OWZP detection scheme uses a diagnostic parameter, the OWZ, and other large-scale environmental variables to identify stronger vorticity regions with weaker deformation forces favorable for TC formation (see the Methods section for further information). This detection scheme is a "phenomenon-based" scheme rather than the more traditional tracking schemes that use specified thresholds of wind speed and vorticity. The OWZP tracking scheme has been further developed to detect TC precursor disturbances, namely tropical depressions (TDs; Tory et al., 2018 [39]). This method is employed here to detect both TDs and TCs. Here, we define "TC" as a vortex equivalent to an observed tropical low with a minimum low-level wind speed of 17.5 m s^{-1} (the U.S. definition of a tropical storm). Previous aqua-planet studies using GCMs have employed resolution-dependent TC tracking schemes that detect the features resembling the observed TCs using windspeed-vorticity thresholds [10,31–34,40]. In contrast to these traditional tracking schemes, the TempestExtremes scheme (Ullrich & Zarzycki, 2017, [41]) uses thresholds of surface pressure and warm core strength to detect TCs. The OWZP scheme is more resolution-independent than traditional tracking schemes, as both reanalyses and model outputs are interpolated to a common resolution (1°) before analysis.

A better theoretical understanding of TC formation is important for projections of the effects of a future, warmer climate. Although most studies project a decrease in TC frequency in a future climate (e.g., [14]), a few studies predict an increase in the formation rate (Emanuel, 2013 [42]; Bhatia et al., 2018 [43]). Additionally, there is inconsistency in future projections of whether there is a decrease in the frequency of low-intensity TCs (Emanuel, 2013 [42]). Recent studies focus on the projected changes in the incipient vortices of TC formation known as TC seeds (i.e., weaker vortices) to better understand the differences in the future projections of the frequency of such storms. Vecchi et al. (2019) [44] examine future changes in numbers of the TC seeds in a high-resolution coupled ocean–atmosphere general circulation model, finding an increase in the frequency of TC seeds but with a reduction in the likelihood of development of these seeds to developed storms. In contrast, Sugi et al. (2020) [45], using two different high-resolution atmospheric GCMs, show a decrease in the future TC seed frequency, as do Yamada et al. (2021) [46]. A recent study by Hsieh et al., (2020) [47] notes a varied response in the frequency of TCs and seeds in idealized AGCM experiments using different SST gradients. These differences in the simulations are explained by the changes in the ventilation index (Tang and Emanuel 2012 [48]), suggesting that this index plays a role in explaining the varied responses of TC frequency projections across the models. These discrepancies in the TC seed frequency changes motivate us to use a phenomenon-based TC tracking scheme to observe the changes in TD frequency in different idealized simulations. Observing the response of incipient vortices to increased SSTs using idealized simulations aims to improve our understanding of the relationships between climate and initial seed vortices.

In the current study, we employ the OWZP scheme to detect locations within marsupial pouches that have the potential for TC formation using appropriate thresholds of the large-scale environmental conditions. For this, we analyze aqua-planet simulations using a high-resolution atmospheric climate model with three different values of globally uniform SSTs. We aim to see whether there is a similar response to increased SSTs of TCs detected within pouch locations using a resolution-independent scheme to that of the TCs identified using the resolution-dependent detection schemes in these idealized simulations. We also examine whether TCs detected within marsupial pouches using the phenomenon-based scheme have better relationships to large-scale environmental conditions than the TCs detected using the traditional TC tracking scheme. The comparison of the OWZP scheme with the CSIRO TC detection scheme (Horn et al., 2014) [49] in identifying the TCs in idealized simulations also helps us to identify a more generalized TC detection and tracking scheme that could apply to different climates for different model resolutions.

Section 2 contains a description of the model and the analysis methods, Section 3 details the results, Section 4 provides a discussion and Section 5 a summary of the main results of the study.

2. Materials and Methods

In the current study, we use the Australian Community Climate and Earth-System Simulator GCM (ACCESS 1.3, Bi et al., 2013 [50]). The description of the model and the simulations here mostly follow those of Sharmila et al. (2020) [5] and Walsh et al. (2020) [10]. The ACCESS model uses the UK Met Office Unified Model (UM v8.5) atmospheric part that includes the upgraded parametrization schemes of the Global Atmosphere (GA 6.0) configuration. The model has a horizontal resolution of about 40 km. It also includes the JULES (Joint UK Land Environment Simulator) land surface model (Best et al., 2011 [51]; Walters et al., 2017 [52]). The atmospheric dynamical core uses the semi-implicit, semi-Lagrangian approach in a longitude–latitude grid, solving both the large-scale processes (solar and terrestrial radiation and precipitation) and sub-grid scale processes (boundary layer processes and convection). A prognostic cloud fraction and prognostic condensate (PC2) scheme is employed (Wilson et al., 2008, [53]). The model also uses the radiation scheme of Edwards and Slingo (1996) [54] for representing the cloud droplets and the boundary layer scheme of Lock et al. (2000) [55] with additional modifications for representing the boundary layer, as described in Lock (2001) [56] and Brown et al. (2008) [57]. The model is initialized with the atmospheric conditions of 1st September 1988 which are produced from the spin-up of 63 years from 1st September 1950 with the Atmospheric Modelling Inter-comparison Project (AMIP-II) SSTs (Kanamitsu et al., 2002, [58]).

The ACCESS model idealized simulations of the current study include spatially homogenous SSTs as the lower-boundary conditions with seasonally varying radiative forcing, which creates moderate meridional temperature gradients in the middle and upper troposphere. The chosen constant SST values are 25 °C, 27.5 °C and 30 °C. Additionally, these simulations have the same varying Coriolis parameters with latitude as in a current climate simulation of Earth. We initialize the atmosphere with values of 1 September 1988, discarding the first four months of the simulation as a run-in period and subsequently perform the idealized simulations for five years.

Two TC tracking schemes are employed. We use a traditional TC tracking scheme, the **Commonwealth Scientific and Industrial Research Organization** (CSIRO) scheme (Horn et al., 2014, [49]), which detects the circulations that resemble the observed TCs. In addition, a phenomenon-based TC detection and tracking scheme is employed using the **Okubo–Weiss zeta parameter** (OWZP; [8,38]), which detects the circulations that have the potential for TC formation within a marsupial pouch region using threshold conditions of large-scale environmental variables.

The CSIRO scheme detects TCs using the following criteria (Horn et al., 2014 [49], Walsh et al., 2020 [10]):

- absolute vorticity at 850 hPa higher than a threshold of $1 \times 10^{-5} \text{ s}^{-1}$;
- mean wind speed at 850 hPa within a 700 km box region around the detected center of the storm should be higher than at 300 hPa by at least 3 m s^{-1},
- 10 m wind speed should be greater than the resolution-dependent threshold of 16.5 m s^{-1} [59] for the 40 km resolution model used here. For a storm to be declared a TC, the minimum duration of satisfying the above thresholds is 48 h.

The OWZP scheme (Tory et al., 2013a,b, [38,60]) involves two steps:

- identification of the grid points that satisfy the initial thresholds (Table S1 of the supporting information) and merging nearby grid points into a single clump; and
- tracking these clumps forward in time until there exists no circulation by using the storm position determined by the 700 hPa steering wind. These clump positions along the track are verified by specified core thresholds (Table S1) and are indicated by True (satisfies the thresholds) and False (does not satisfy the thresholds). A particular track is considered as a TC if it satisfies the conditions for a period of 48 h, that is five consecutive True positions (Figure S1). The fifth True position is considered as the TC genesis location (Bell et al., 2018 [61]). The time interval of the data is 12 h.

Tory et al. (2013d) [8] shows only minor changes in the global TC detection numbers in the current climate by keeping the specific humidity threshold to a value of 0 g/kg instead of 14 g/kg. Therefore, the specific humidity threshold of the OWZP scheme here is taken as zero since for increasing values of the SST as in the experiments analyzed here, we observe a significant change in the specific humidity, which leads to differences in the TC detection numbers. We further employ the OWZP scheme to detect the TDs in all three aqua-planet simulations by using similar thresholds of the large-scale variables but with a reduction in the duration threshold to 24 h (Tory et al., 2018) [39]. Tory et al. (2018) [39] tune the OWZP scheme with a tropical disturbance dataset across the Australian region to identify the initial TDs. As all TCs were once TDs, we obtain the non-developing depressions by subtracting the TCs from the total number of TDs. Raavi and Walsh (2020a) [3] give a more detailed description of the tracking schemes.

Here, we consider different environmental variables that were found to act as influencing variables for TC formation in a previous study with a reanalysis dataset [62] and the same set of aqua-planet simulations (Walsh et al., 2020 [10]). We also consider the saturation deficit, one of the components of the ventilation index (Tang & Emanuel, 2012 [48]), an important predictor for TC formation which scales with the changing climatic conditions. Variables examined include the (i) 700 hPa relative humidity (RH; %), (ii) the difference in the potential temperature between the 925 and 200 hPa levels (an indicator of vertical static stability), (iii) 700 hPa saturation deficit, that is the difference between saturation-specific humidity and specific humidity, (iv) the magnitude of vector wind difference between the 850 and 200 hPa levels, the vertical wind shear (VWS; m/s), (v) 500 hPa omega (Pa/s), (vi) 10 m wind speed (denoted "WIND"), and (vii) the 850 and 500 hPa Okubo–Weiss (OW) parameter. We take 12-degree box averages for RH, static stability, saturation deficit, and VWS and 4-degree box averages for omega, WIND, and the OW parameter around the circulation center at the time of the TD genesis for both developing and non-developing cases. Figure S1 shows that TD genesis occurs after three "True" detections, defined as Day 0. This time is 24 h before TC genesis and is chosen to determine whether there are any differences in environmental variables that determine whether a TD goes on to become a TC or not. We use the box-averaged instantaneous values of the environmental variables on Day 0 to plot the histograms. Additionally, we perform a binary classification for the TCs in the SST25 and SST30 experiments to identify the major contributing variables to TC formation under increasing SSTs. Here, we use a C4.5 algorithm for the binary classification, which uses a specified metric called "information gain" that divides the TCs of the SST25 and TCs of SST30 experiments into either of the two classes, with the background environmental conditions as the predictors (Quinlan, 1987 [63]). We remove one of the highly collinear variables (i.e., having correlation greater than 0.5) before performing the classification. For instance, as the omega and OW variables correlate, we retain OW and remove omega for building the decision trees. Additionally, the OW parameter at 850 hPa correlates with the OW at 500 hPa level, so we retain OW at 850 for the classification analysis. The information gain is a probabilistic measure of the decrease in the uncertainty in the dataset for developing cases in both "TCs (SST25)" and "TCs (SST30)" experiments using large-scale environmental variables as predictors. Here, the variable that has the maximum information gain most effectively partitions the data into TCs in SST25 and TCs in SST30 at each node of the decision tree. The construction of the decision tree involves the selection of an environmental variable that has maximum information gain as the "root node." The subsequent splitting of the decision tree stops when it reaches the specified minimum leaf size threshold. The minimum leaf size represents the size of the decision tree; a smaller tree better generalizes the data. Further details of the decision trees using the C4.5 algorithm are given in the Appendix A.

3. Results

3.1. Mean Annual Frequency

Figure 1 shows that the OWZP scheme detects more storms with lower box-averaged maximum 10 m wind speeds (i.e., lower than 10 m/s) compared to the CSIRO scheme (note that 10 m wind speed is not used as a detection criterion in the OWZP scheme and the OWZP values in Figure 1 are calculated after detection is performed). Since the resolution-dependent 10 m wind-speed criteria in the CSIRO scheme eliminates the TC detections below a threshold of 16.5 m s^{-1} instantaneous maximum wind speed at a grid point, we observe fewer storms with box-averaged wind speeds less than 10 m/s. In contrast, the OWZP scheme uses the thresholds of large-scale environmental conditions and does not use any thresholds of the wind speeds. Additionally, since the OWZP scheme is tuned using low-resolution reanalysis datasets, we speculate that modified thresholds obtained after tuning using high-resolution reanalysis could remove some of these detections with weaker intensities. In this study, we remove the OWZP storms below a limit of box-averaged lifetime maximum 10 m/s wind speed for further analysis of TCs (i.e., we used the numbers in data column three of Table 1) to remove the secondary peak in the bi-modal distribution of intensities shown in Figure 1a and focus on the arguably more realistic, more intense storms. In addition, Figure 1 shows an increase in the number of intense storms with increasing SSTs using both the OWZP and CSIRO schemes. We further estimate the mean annual number of OWZP detected storms for each of the constant SST experiments (Table 1). The SST25 experiment is considered as the control experiment, for which we found a global mean number of 3218 TCs. By increasing the temperatures for the SST27.5 and SST30 experiments, we observe a decrease in the global mean annual TC detections, with 2929 and 2923, respectively. Similarly, using the CSIRO scheme, we found 2277 TCs for the SST25 experiment and decreasing numbers for the SST27.5 and SST30 runs, with 1808 and 1586, respectively. The increased detections in all the three constant SST experiments using the OWZP scheme compared to the CSIRO scheme are like those found in a current climate simulation using the ACCESS 1.3 model, which also gives higher detections using the OWZP scheme (Raavi and Walsh 2020a [3]). Using the OWZP scheme (column 3 of

Table 1), we observe a decrease in the frequency of TCs by 12% when increasing the SST from 25 to 27.5 °C and a 15% reduction when increasing the SST from 25 to 30 °C. In contrast, the CSIRO scheme has a more significant decrease of 21% and 30% by increasing the SSTs from 25 to 27.5 and 30 °C, respectively. Overall, we observe a reduction in the genesis frequency with increasing SSTs using both tracking schemes.

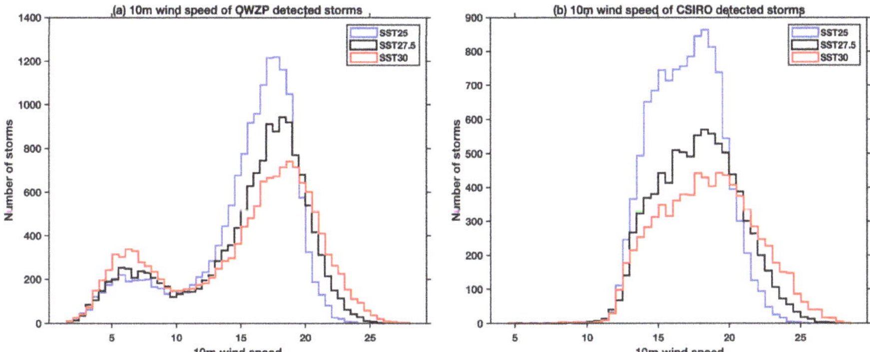

Figure 1. The distribution numbers of cyclones with maximum 10 m wind speed (x axis) attained during the lifetime of each of the storms detected by both OWZP and CSIRO tracking schemes for the three aqua-planet experiments for five years of simulation. Wind speed is averaged over a box 2° on a side, centered around the TC.

Table 1. The global mean annual number of TCs and TDs for each of the aqua-planet simulations using the OWZP and CSIRO detection schemes.

Experiment	OWZP Scheme (TC)	CSIRO Scheme (TC)	OWZP (Greater Than 10 m/s)	OWZP Scheme (TDs)	Non-Developing Cases (ND)
SST25	3218	2277	2583	5007	1789
SST27.5	2929	1808	2286	4449	1520
SST30	2923	1586	2197	4463	1540

The OWZP scheme appears to be better at detecting weaker systems at formation than the CSIRO scheme. Thus, storms in the OWZP scheme generally appear to have longer lifetimes (see also Figure 6). As a result, there is an increased number of TC detections (Figure 1) in all three constant SST experiments compared to the CSIRO scheme.

3.2. Geographical Distribution of TC Genesis Frequency

Figure 2 shows the geographical distribution of TC genesis per 2-degree latitude box across the globe using both the CSIRO and OWZP schemes in the three different SST experiments. This is calculated using a multiplicative factor to convert the TC genesis originally detected in 2-degree latitude by 2-degree longitude boxes into 2-degree latitude boxes, which have equal areas over the entire globe. The aqua-planet geographical distribution of TC genesis differs compared to the current terrestrial climate simulation (Figure 1 of Raavi and Walsh, 2020a [3]) due to the absence of land with favorable SSTs everywhere and weaker vertical wind shear regions due to the lack of strong meridional temperature gradients. In the OWZP detection scheme, we observe fewer detections between 0 to 15° latitudes with increased detections poleward of 15° in each of the three experiments. In contrast, the CSIRO detection scheme has fewer detections between and 0 to 25° latitudes and increased detections poleward of 25°. The decrease in TC genesis density with increasing SSTs is apparent in both the schemes. Additionally, the symmetric SST specification in both hemispheres produced a similar number of detections in both hemispheres.

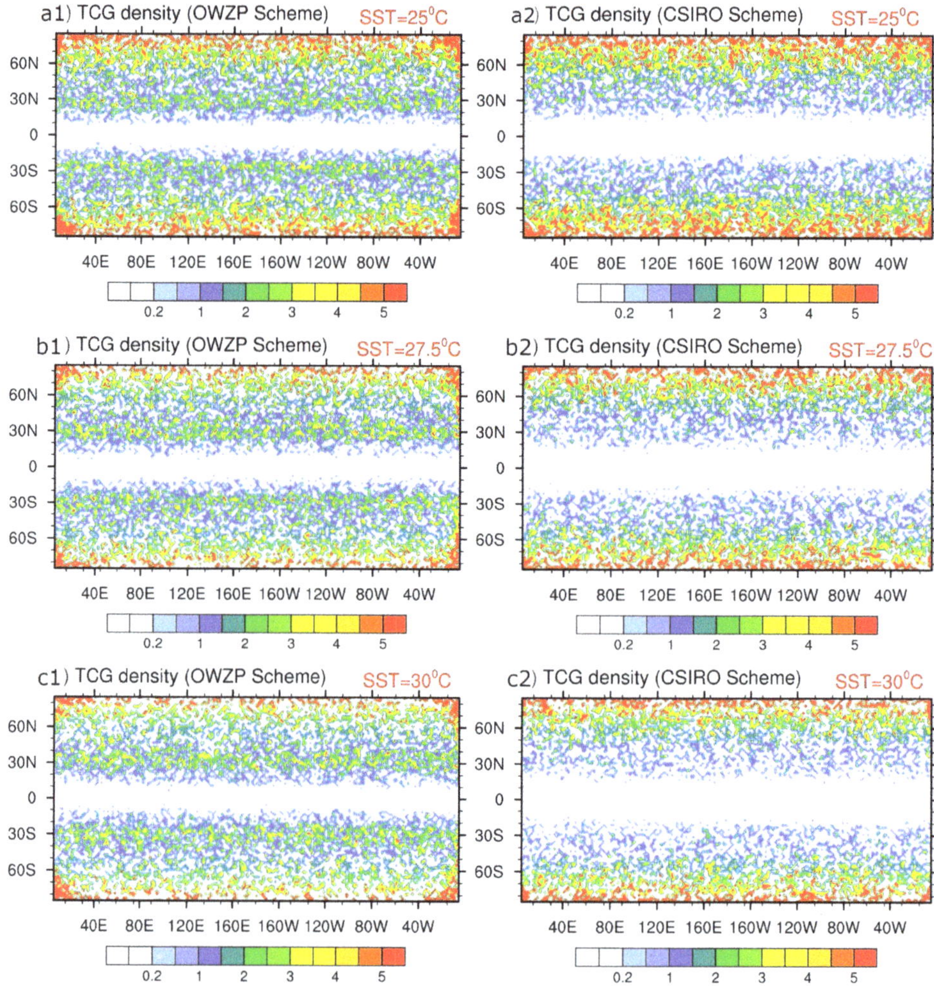

Figure 2. Spatial distribution of number of TCs forming per 2-degree latitude box per year for the three constant SST experiments using both OWZP and CSIRO schemes. The color bar indicates the number of genesis events per ° 2-degree latitude box region.

From the spatial distribution of TC formation, we observe individual differences between the two schemes in specific latitude bands. Figure 3 shows the latitudinal distribution of the TC genesis frequency using both the schemes, showing differences between them in the latitudes 10 to 45°. Using the OWZP scheme, there is a gradually increasing TC frequency with latitude between 10 to 25° and a decrease with latitude in the 30 to 45° band, whereas using the CSIRO scheme, detections generally increase poleward of 20° in all three experiments. The difference between the CSIRO and the OWZP scheme is mainly due to the peak at latitude 30° in the OWZP scheme. When examining the 10 m wind speeds for different latitude bands, we notice that the OWZP-detected storms in the 0–30° latitude band have a lower 10 m wind speed magnitude compared to the CSIRO storms (not shown), which are detected using a resolution-dependent wind speed threshold of 16.5 m/s. Figure 4 shows the mean annual TC frequency across different latitude bands for all three experiments using both schemes. Note that Figure 4 shows a different metric to Figure 3, as Figure 4 gives absolute numbers of formation, while Figure 3 gives values per 2-degree latitude box. Figure 4 shows that while TC formation decreases with increased SST at all latitudes in the CSIRO tracking scheme, in the OWZP scheme, formation decreases a little in the 0–25° latitude band, increases from 30–40°, and then decreases further poleward. A likely explanation is due to the differences in the tracking schemes: the OWZP scheme detects TCs earlier in their lifecycle, thus giving maximum formation in the subtropics, and it is clear from Figure 4 and Figure S5 that there is a poleward movement of this maximum with increased SST, as while the typical latitude of maximum formation is between 25 and 30 degrees for the SST25 experiment, it is between 30 and 35 degrees for the higher SST experiments. One possible explanation for this result is given by Sharmila and Walsh (2018) [64], who observe that a poleward shift in the TC genesis locations is due to the changes in the Hadley circulation. Examination of the mass stream function in these experiments (Figure S2) shows a clear three-cell circulation, like the terrestrial climate but considerably weaker (see, for instance, Peixoto and Oort, 1992 [65]; their Figure 7.19). Further documentation of the mean state of the simulations is contained in the supplementary material (Figures S3 and S4 and accompanying text). We speculate that this simulated Hadley circulation is driven by the meridional gradient of radiation that is included in these experiments, but without a surface meridional gradient of temperature, the resulting Hadley circulation is much weaker than observed. Alternatively, this poleward movement of TC formation is also consistent with the theory of Chavas and Reed (2019) [17], as in the current experiments there is a slight poleward movement in the latitude of maximum potential intensity with increased SST (not shown). These differences in the detections between latitude bands with OWZP and CSIRO schemes are perhaps due to fundamental differences in the tracking schemes (see the Discussion section for more on this point). Note that baroclinic disturbances in these simulations are almost non-existent due to the absence of substantial meridional temperature gradients.

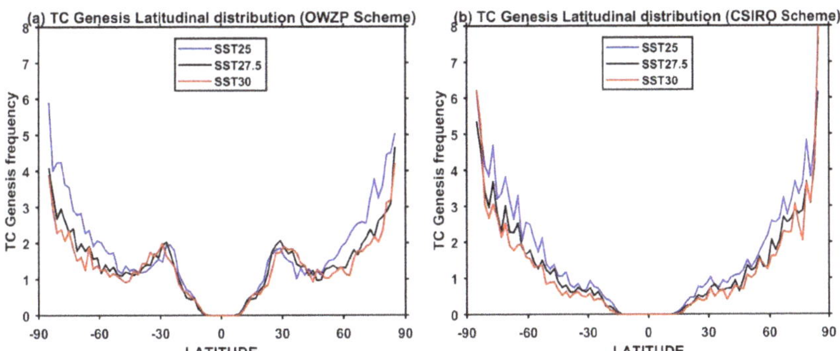

Figure 3. Latitudinal distribution of genesis frequency per 2-degree latitude box for the three constant SST experiments using (**a**) the OWZP and (**b**) CSIRO detection schemes.

Figure 5 shows the latitudinal variation of mean environmental parameters for comparison with Figures 3 and 4. The comparison here focuses on the OWZP detection scheme (Figure 4a). In general, like formation, the distributions of the environmental parameters appear to be moving poleward with increased SST. The static stability (Figure 5a) is greater at all latitudes in the SST30 and SST27.5 experiment than in the SST25 experiment. This would explain the lower formation in the SST30 experiment poleward of about 45° but would not explain the increased genesis in the subtropical region. Relative humidity (Figure 5b) is lower for higher SST at all latitudes but particularly between 0 and 30°. Similar results can be seen for saturation deficit (not shown). Vertical wind shear (Figure 5c) shows somewhat lower values in the SST30 experiment between about 5° and 35°, in agreement with the increase in formation in this region. Conversely, higher wind shear with increased SST is seen poleward of 35° to about 60°, in agreement with lower formation in this latitude band. Poleward of 60°, lower formation rates are not explained by wind shear changes. No clear latitudinal patterns were seen for differences in the omega or Okubo–Weiss parameter (not shown).

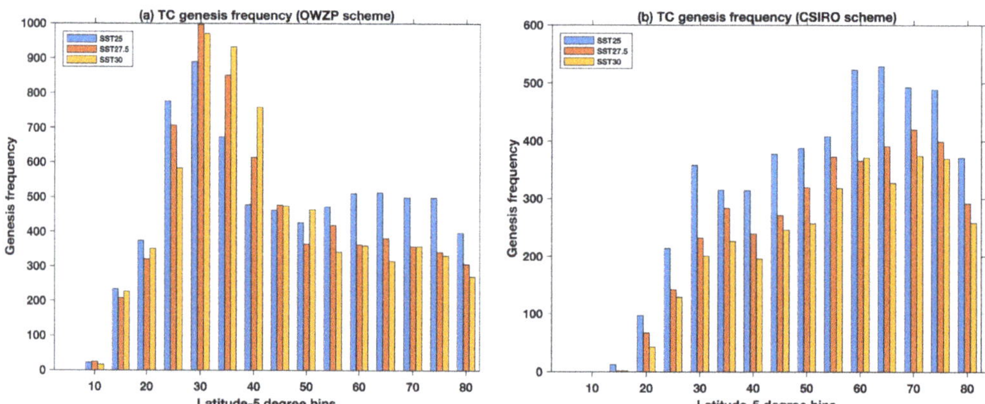

Figure 4. Mean annual frequency of detected genesis of storms using (**a**) the OWZP detection scheme and (**b**) the CSIRO scheme for each 5° latitude band.

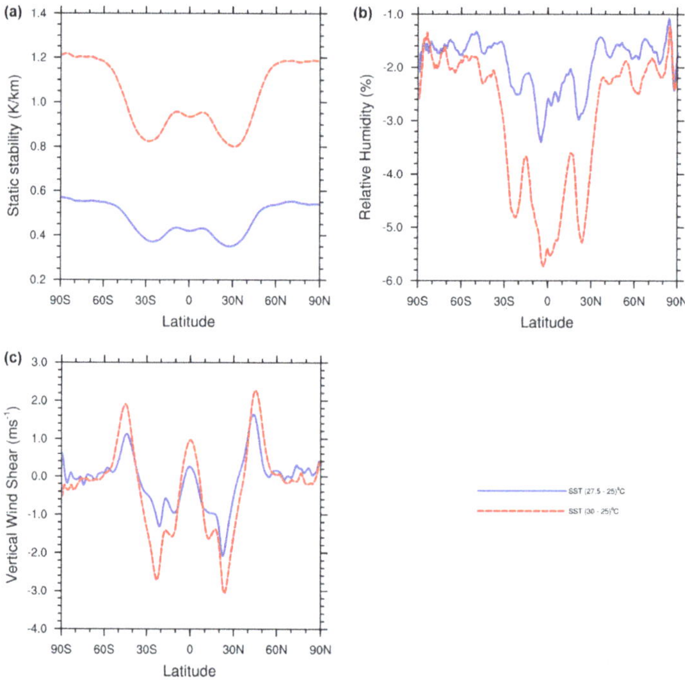

Figure 5. Mean values of differences in environmental parameters between experiments, for comparison with Figures 3 and 4, for (**a**) static stability (K km^{-1}); (**b**) relative humidity (%); (**c**) vertical wind shear (m s^{-1}) for each of the SST experiments.

3.3. TC Lifetime and Tracks

The OWZP scheme has a higher number of storms with a more prolonged duration, whereas the CSIRO scheme gives shorter-duration storms. This can be seen in Figure 6 and Figure S5. The lifetime of TCs in the same model forced with interannually varying AMIPII SSTs (Kanamitsu et al., 2002 [58]) for the period 1990–2009 shows that the CSIRO scheme-detected storms have a shorter duration compared to the OWZP detected storms (Raavi and Walsh, 2020a [3]). Raavi and Walsh (2020a, [3]) note that the CSIRO scheme-detected TC tracks are a subset of the OWZP scheme-detected tracks. To identify whether the CSIRO tracks are a subset of the OWZP tracks in these aqua-planet simulations, we have considered a particular month (August) in the year 1990 for the SST25 experiment and plotted individual tracks of TCs detected by both the schemes. Figure 6 shows that CSIRO tracks are a subset of OWZP tracks for the SST25 simulation with shorter tracks. Similar results are observed in the other two constant SST experiments (not shown). The TC tracks in these simulations show that genesis from the OWZP scheme occurs mostly in the subtropics, and in both tracking schemes, the TCs move poleward and westward. The poleward movement of the TCs is mainly due to the beta-drift mechanism caused by the presence of the latitudinally varying Coriolis force in these simulations [66–68].

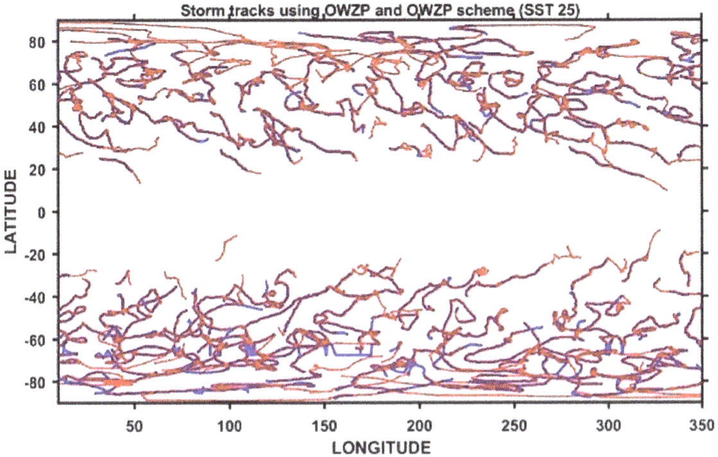

Figure 6. Tracks of the storms detected using the OWZP scheme (red lines) and CSIRO scheme (blue lines) for August 1990 of the SST 25 experiment.

3.4. Changes in the Large-Scale Environmental Conditions for the Constant SST Experiments

To understand the differences in the genesis frequency between different latitude bands, we have calculated the large-scale environmental conditions at the time of the TD genesis in developing cases (Day 0, as shown in Figure S1) for the OWZP storms for each of the latitude bands. Figures 7 and 8 show the mean environmental conditions around the TD center for all three SST experiments. These plots give insight into the typical environment near the center of developing TDs for varying latitudes and SST values. Of all the variables, we observe that the saturation deficit and stability parameters show the most significant differences (greater than 95% significance level using two sample Student's *t*-tests) between the three simulations. That is, with an increase in SST, we observe an increase in the stability of the atmosphere (Figure 7). The stability values also show an increasing trend with latitude, with the polar regions having higher values compared to the other latitude bands in all the three experiments. This rising trend of stability with latitude is accompanied by increasing precipitation towards poles in these sets of simulations, as also shown in Figure 3 of Walsh et al. (2020) [10]. We speculate that higher values of precipitation lead to the release of the latent heat flux during the condensation process that increases the temperatures of the upper atmosphere, thereby stabilizing the atmospheric column. This increased stability with increased upper tropospheric temperature leads to a decrease in the frequency of storms at higher latitudes. Similarly, the saturation deficit increases with increasing temperatures across all the latitude bands. In contrast, Figure 7 only shows prominent differences in RH between about 0° and 40° latitude, with the SST25 experiment having higher relative humidity compared to the SST27.5 and SST30 experiments. Therefore, with increasing temperature, there is a global increase in the stability and saturation deficit of the atmosphere, both of which are more unfavorable for TC formation.

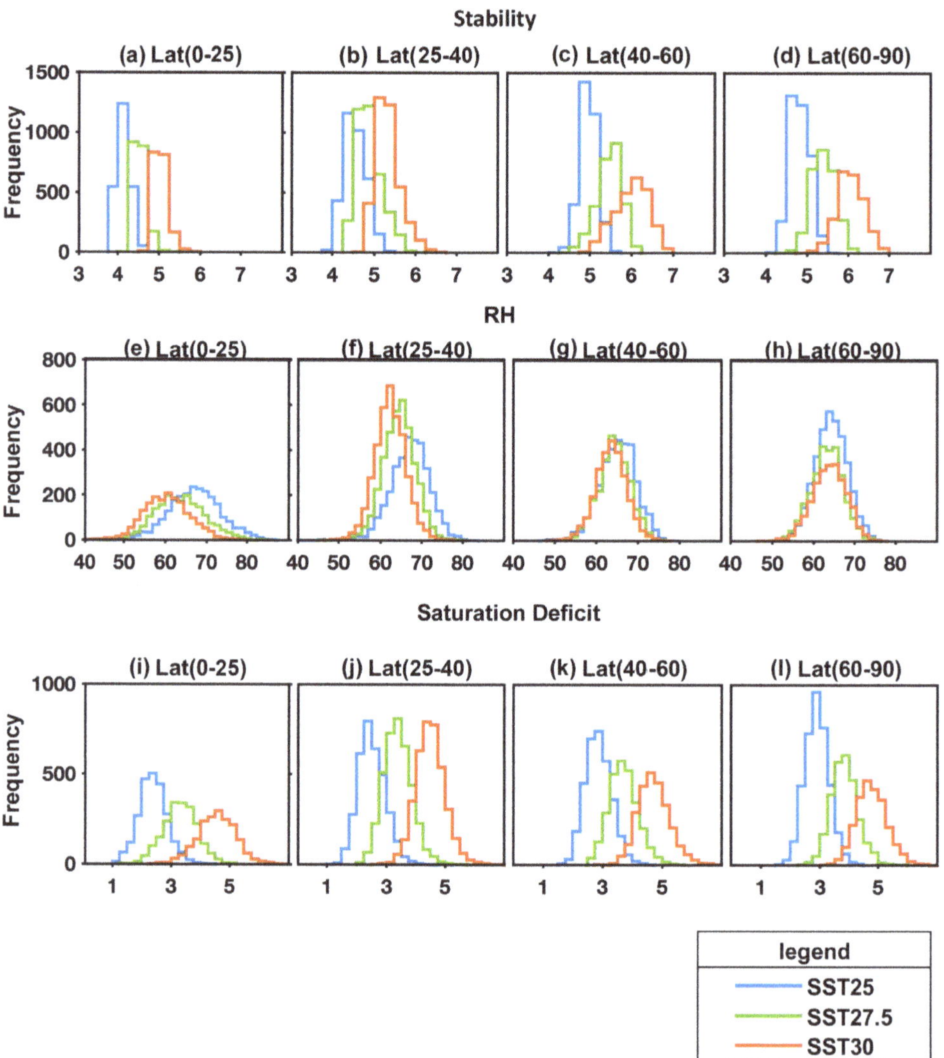

Figure 7. Histograms of mean values of stability parameter (**a–d**; K/km); RH (**e–h**; %) and saturation deficit (**i–l**; kg/kg) for OWZP-detected developing TCs at Day 0 (TD genesis in developing cases) within a 12-degree box region around the circulation center in all three constant SST experiments.

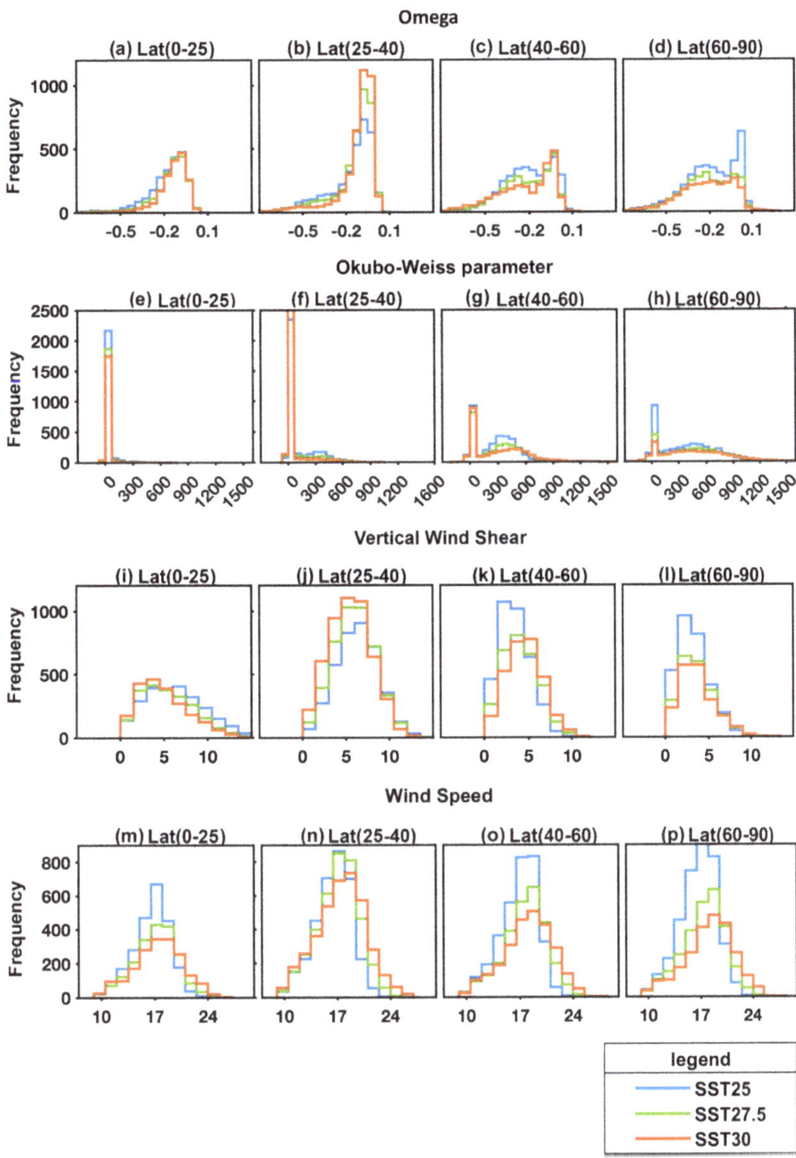

Figure 8. The same as Figure 7 for omega (**a–d**; Pa/s) and 850 hPa OW parameter (**e–h**; s$^{-2} \times 10^{-1\circ}$) within a 4-degree box region around the circulation center at Day 0; vertical wind shear (**i–l**; m/s) and wind speed for OWZP detected storms (**m–p**; m/s) within a 12-degree and 4-degree box average, respectively at Day 0 across latitude bands in all three constant SST experiments.

Nevertheless, other variables appear to show more agreement with the latitudinal variation in TC formation shown in Figure 4. For instance, Figure 8 shows that for developing TDs poleward of 55°, omega values are generally more negative (higher upward mass flux) for increased SSTs, Okubo–Weiss parameter values are generally more positive, and there is greater wind shear, while Figure 4 shows that at these latitudes, there is lower TC formation for higher SSTs. The 850 hPa OW parameter shows a similar latitudinal distribution with increased SST to that of omega. There is also a gradually increasing trend in its magnitude with increasing latitude. This implies that TDs at genesis in the low latitudes have lower values of the OW parameter compared to the storms in higher latitudes. The higher values of the OW at the higher latitudes (defined here as poleward of 40 degrees latitude) have enhanced vertical stability of the atmosphere compared to the lower latitudes, indicating that as stability increases, the vortices need to have stronger vorticity for genesis to occur. This also means that the vortices that have weaker strength, indicated by weaker values of the low-deformation vorticity, do not develop into TCs, and may become non-developing cases. Therefore, there may be a reduction in the formation of the weaker strength vortices in the more stable atmospheric conditions.

The other variables considered are the VWS and wind speed (Figure 8); we notice in the latitude band 25–40°, SST30 and SST27.5 have slightly lower values of shear compared to SST25. These lower values of vertical wind shear led to increased detections in the SST27.5 and SST30 experiments compared to the SST25 (see Figure 4). Similarly to the results of Walsh et al. (2020) [10], we observe that the VWS at lower latitudes increases with a decrease in the SSTs. This is due to the presence of meridional temperature gradients aloft generated by retaining Earth-like radiative forcing in the current simulations, rather than imposing globally constant insolation (for more explanation of this point, see Walsh et al., 2020 [10]).

3.5. Classification of the TCs in the SST25 and SST30 Experiments

Although the distributions indicate that as the stability and the saturation deficit increase, the TC frequency decreases, it is still not clear what other environmental conditions lead to TC formation under increased SSTs in different latitude bands. Therefore, we have performed a binary classification analysis for the TCs in the SST25 and SST30 experiments using a C4.5 algorithm to find out the combination of the environmental conditions favoring TC formation under changing environmental conditions due to increased SSTs. This algorithm identifies the most prominent parameters needed to classify the SST25 and SST30 TCs given different background environmental parameters as the predictors. Table 2 shows the different decision rules obtained from the classification algorithm. To improve the statistical power of the method, we have grouped the data into larger latitude bands. For example, in the latitude band of 0–25 degrees, the most statistically important rule (rule 1) is physically related to higher relative humidity in the SST25 experiment. Thus, in this latitude band, relative humidity is a better discriminator of TC formation in the SST25 experiment than in the SST30 experiment. These data are then excluded from subsequent analysis, so rule 2 analyses lower relative humidity values, finding that for those data the most important physical variable for TC formation is wind shear less than 5.7 m s^{-1} in the SST30 experiment. This rule favors the SST30 data because relative humidity is lower in that experiment. Thus, rule 2 does not mean that lower relative humidity promotes TC formation. In the latitude band 25–40°, rule 1 also indicates that higher values of relative humidity led to TC formation in the SST25 experiment. Rule 2 shows that favorable values of the OW parameter are required for TC formation in the SST30 experiment, while rule 4 adds lower values of vertical wind shear. Rule 3 indicates that in the SST25 experiment, favorable values of OW and moderate values of relative humidity are required for formation. It is important to note that these rules also imply that higher values of OW and lower values of vertical wind shear are required for a TC to form in the SST30 experiment than in the SST25 experiment. In the latitude band 40–60°, rule 1 shows that higher values of the OW parameter are required in the SST30 experiment, and our analysis also showed that higher OW is correlated with higher upward mass flux (more negative values of omega). Rules 2 and 3 favor SST25 because OW values are typically lower in that experiment. For these lower values, higher mid-level relative humidity and lower vertical wind shear are required for formation. Rule 4 indicates that for lower relative humidity values, higher OW is required for formation in the SST30 experiment than in the SST25 experiment. Finally, in the latitude band 60–90°, rule 1 indicates that higher wind speeds are required for TC formation in the SST30 experiment, since in that experiment, there are generally less-favorable thermodynamical conditions, such as a higher stability and saturation deficit of the atmosphere. Rules 2 and 3 show that for weaker vortices, more favorable values of wind shear are required. Rule 4 shows that when wind speeds are lower, there needs to be favorable shear for the TC formation in the SST30 experiment. Overall, this analysis indicates that under more stable atmospheric conditions, incipient vortices with higher low-deformation vorticity (OW), higher surface wind speeds, and higher values of upward mass flux led to TC formation in the higher latitudes. In the lower latitudes where the atmosphere is stable but to a lesser extent compared to the higher latitudes, lower values of vertical wind shear favor TC formation in the SST30 experiment.

Table 2. Decision tree rules obtained after applying the C4.5 classification algorithm to the SST25 and SST30 experiments for different latitude bands.

SST25 and SST30 (0–25)
1. if RH > 65.43 then TC forms in SST25
2. if RH $\leq$ 65.43 and VWS $\leq$ 5.7 then TC forms in SST30

SST25 and SST30 (25–40)
1. if RH > 67 then TC forms in SST25
2. RH $\leq$ 67 and 10 < OW $\leq$ 70 then TC forms in SST30
3. if RH $\leq$ 67, OW > 70, and RH > 63 then TC forms in SST25
4. RH $\leq$ 67, OW > 70, RH < 63, and VWS < 6 then TC forms in SST30

SST25 and SST30 (40–60)
1. if OW > 568 then TC forms in SST30
2. if OW < 568, VWS < 4.3 then TC forms in SST25
3. if OW < 568, VWS >4.3, and RH > 66 then TC forms in SST25
4. if OW < 568, VWS >4.3, RH < 66, and OW > 353 then TC forms in SST30

SST25 and SST30 (60–90)
1. if WIND > 18 then TC forms in SST30
2. if WIND $\leq$ 18 then VWS $\leq$ 2.1 then TC forms in SST25
3. if WIND $\leq$ 18, VWS > 2.1, and VWS $\leq$ 6.5 then TC forms in SST25
4. if WIND $\leq$ 18, 6.5 < VWS $\leq$ 12 then TC forms in SST30

3.6. Developing Versus Non-Developing Tropical Depressions

To understand whether there is an overall decrease in formation activity or whether there is an increase in the non-developing cases compared to the developing cases with increasing global temperatures, we use the OWZP scheme to detect both developing and non-developing TDs (Tory et al., 2018) [39]. Table 1 shows the global annual frequency of the total TDs, including developing and non-developing depressions. We note that there is a decrease in the global mean annual frequency of TDs and non-developing TDs with increasing SST, like the trend for developing cases, although this is latitude-dependent (see Figure 9). This reduction in the initial TDs may imply that there will be a reduction in the number of weaker storms in a future climate. Although the intensity of the storms (as measured by their 10 m wind speeds) is weaker in the ACCESS model compared to the observations (Figure 1), there is an increase in the frequency of intense storms with increasing temperatures, for both OWZP and CSIRO storms. Using two high-resolution atmospheric models, Sugi et al. (2020) [45] noted that the frequency of TC seeds (i.e., weaker pre-storm vortices) decreases with increasing temperatures, with the seeds detected based on the 850 hPa vorticity and a warm core criterion (the geopotential height difference between 200–500 hPa). Figure 9 shows that in the latitude bands 0–25° and 25–50°, there is an increase in the mean number of non-developing cases with increasing temperatures. In contrast, at latitudes poleward of 50°, we observe a decrease in the non-developing cases with an increase in the SSTs. To further understand the reasons for the changes in the TC frequency, for both developing and non-developing cases in all three experiments across different latitude bands, we evaluate the changes in large-scale variables due to different underlying SSTs.

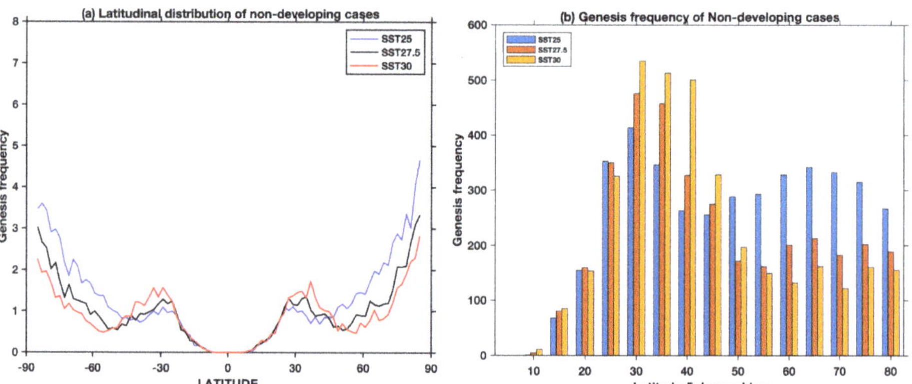

Figure 9. Latitudinal distribution of genesis frequency of non-developing depressions at Day 0, detected using the OWZP detection scheme for the three constant SST experiments: (**a**) distribution per 2-degree latitude box; and (**b**) in 5° latitude bands.

3.7. Changes in the Large-Scale Environmental Conditions for Both Developing and Non-Developing Circulations

To show more clearly the spread of values of the various environmental variables at the time of TD genesis, we plot the histograms of these variables. Figure 10 shows the variations in the stability of the atmosphere for the SST25 and SST30 experiments for both developing and non-developing cases (for clarity, the intermediate SST27.5 experiment is omitted). A similar pattern exists for both developing and non-developing cases: there is increasing stability with increasing SSTs. The increasing stability leads to a net decrease in the formation of the TDs with increasing SSTs for both developing and non-developing cases. The saturation deficit parameter also shows similar differences between developing and non-developing cases in that there is an increase in the saturation deficit with increasing temperature. The other thermodynamical variable is RH, which shows that there is a decrease in the RH with increasing temperatures in both developing and non-developing cases. The non-developing cases have lower mid-level relative humidity compared to the developing cases in all three constant SST experiments. The increasing number of non-developing cases with increasing temperatures in the latitudes 0–45° (Figure 9) is perhaps due to reduced mid-level relative humidity, increased saturation deficit and increased stability of the atmosphere with rising temperatures.

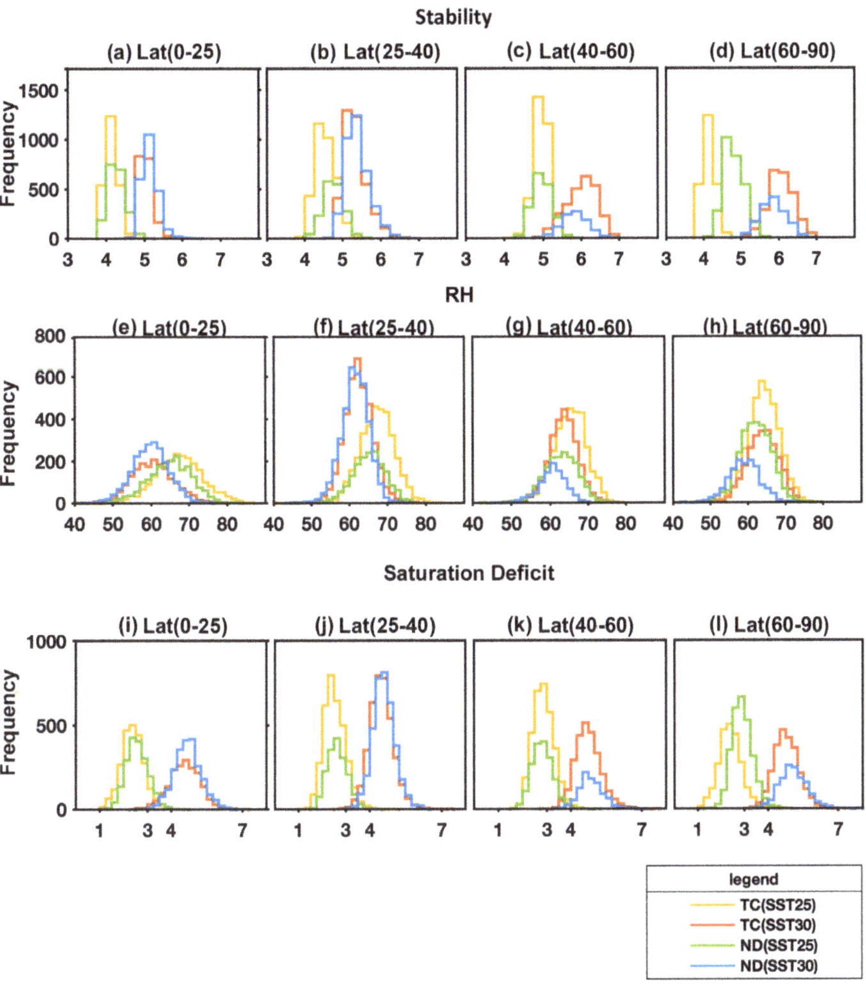

Figure 10. Histograms of stability parameter (**a–d**; K/km); RH (**e–h**; %) and saturation deficit (**i–l**; g/kg) for both developing (TC) and non-developing (ND) tropical depressions at Day 0 (TD genesis) within a 12-degree box region around the circulation center across different latitude bands in the SST25 and SST30 experiments.

Previous works have shown that increased values of omega (i.e., reduced upward mass flux) lead to decreased TC frequency in future projections using GCMs due to weaker atmospheric vertical circulation caused by a decreasing lapse rate with increasing temperatures (Sugi et al., 2002 [69], 2012 [70]). To show the differences in circulation between developing and non-developing storms, we group the circulation variables at TD genesis for both developing and non-developing TDs. The differences in the circulation variables between developing and non-developing disturbances are considerably more pronounced than shown in Figure 10. Here, Figure 11 shows that developing cases have more negative values of omega (higher upward mass flux), favoring formation compared to the non-developing cases. The distributions of OW and 10 m wind speed variables have a similar distribution to omega, with developing cases having more favorable values with stronger circulation compared to the non-developing cases. These results suggest that initial vortices with stronger low-deformation vorticity and higher upward mass flux and 10 m wind speeds lead to the formation of a TC. The non-developing cases have narrower distributions that are typically clustered closer to zero than the developing cases, indicating a considerably weaker circulation in the ND cases. The distributions of the vertical wind shear indicate roughly similar distributions for developing and non-developing disturbances.

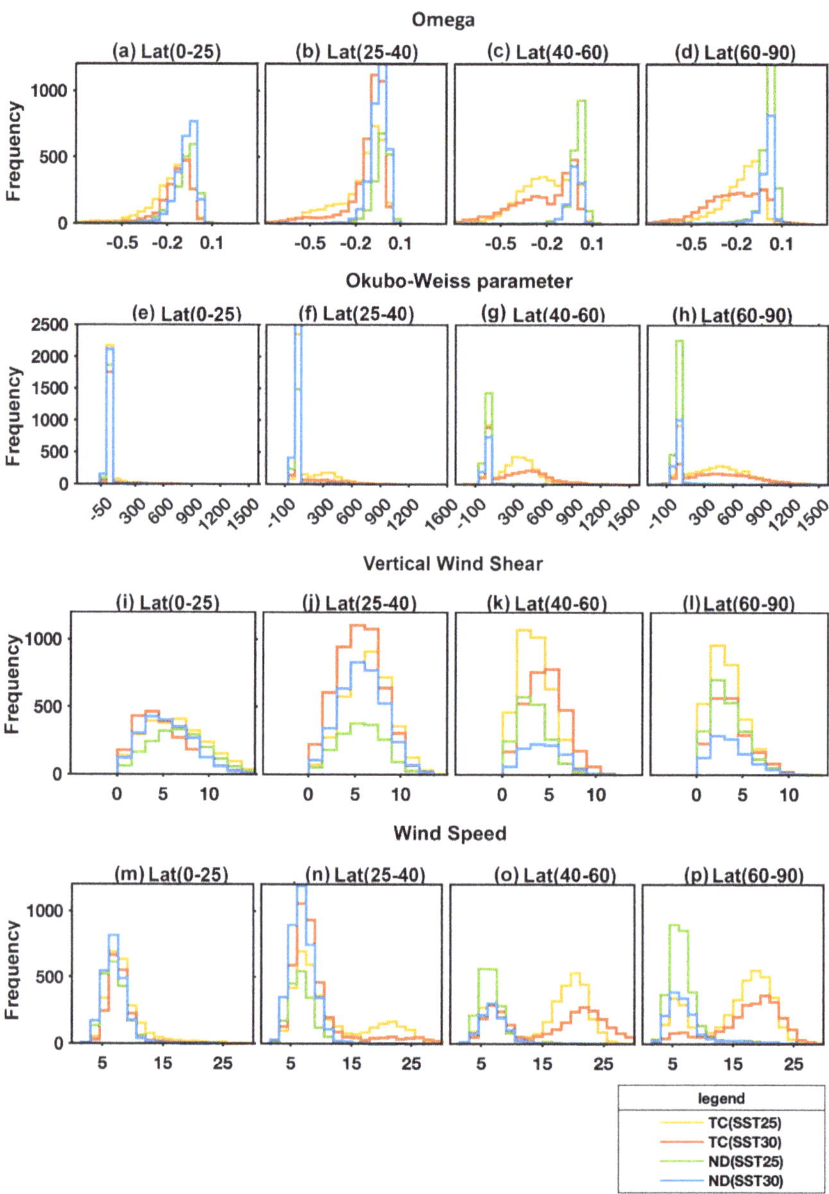

Figure 11. Histograms of omega (**a–d**; Pa/s) and OW parameters (**e–h**; $s^{-2} \times 10^{-10}$) within a 4-degree box region around the circulation center at Day 0; vertical wind shear (**i–l**; m/s) and wind speed (**m–p**; m/s) for 12-degree and 4-degree box averages, respectively around the circulation center at Day 0 for both developing and non-developing tropical depressions across different latitude bands in the SST25 and SST30 experiments.

This analysis of the environmental differences between developing and non-developing tropical depressions shows that the RH, OW, and omega explain the differences in the developing and non-developing cases, given the increased saturation deficit and stability of the atmosphere with increasing SSTs. These results agree with both Sugi et al. (2002) [69] and Emanuel (2013) [42], showing that the decrease in TC frequency is due to the decrease in the upward mass flux and reduced mid-level relative humidity content. Therefore, unfavorable environmental conditions such as lower RH content, lower values of upward mass flux, and weaker circulation (indicated by lower OW values) lead to the non-development of the circulation.

4. Discussion

To understand the environmental conditions and the associated mechanisms leading to the changes in TC genesis frequency within marsupial pouches, we analyze idealized climate model simulations with different values of globally uniform SSTs in a high-resolution atmospheric model (ACCESS). Here, we compare the TC frequency changes within the model obtained from two fundamentally different TC tracking schemes. Both schemes show a decrease in the global TC genesis frequency with increasing temperatures, albeit with a difference in the magnitude of the decrease. The OWZP scheme shows a mix of increases and decreases with latitude and SST. While the simulated climate in these experiments has several differences from the observed terrestrial climate, the general result of a global decrease in TC numbers with increased temperature is consistent with the predicted response of the terrestrial climate to global warming simulated by most climate models (Walsh et al., 2016 [16]; Knutson et al., 2020 [14]). Most of this decrease in the simulations appears to occur at higher latitudes, however, where the TC climatology is very non-terrestrial. Additionally, Figure 1 shows that there is a modest increase in numbers of more intense TCs with SST, accompanied by a decrease in the overall genesis frequency. Knutson et al. (2015) [71] showed similar results, although with a much finer resolution model and a better simulation of intensity.

The TC frequency per 2-degree latitude box gradually increases from the subtropics to the poles using both tracking schemes, with some differences at specific latitudes. Merlis et al. (2016) [33] also noted that there is an increase in the TC frequency towards the poles in a constant SST GCM simulation. When comparing the overall TC detections in different latitude bands, the OWZP scheme has a higher number of TCs in the tropics (Figure 4) compared to the CSIRO scheme. As the resolution-dependent CSIRO scheme uses a threshold of 16.5 m s^{-1} instantaneous grid point wind speed to remove TCs below this threshold, we observe fewer TCs using this scheme compared to the OWZP scheme. The reduced number of TCs using the CSIRO scheme across this 0–25° latitude band is due to the higher 10 m wind speed threshold, whose value in the CSIRO scheme is set to be consistent with the resolution of the model. There are a higher number of detections in mid-latitudes and poles compared to the terrestrial climate simulations using both the detection schemes, perhaps due to more favorable conditions for TC formation within the aqua-planet simulations. On examining the changes in the genesis frequency between different SST experiments, both the OWZP and the CSIRO show a decrease in genesis at 0–25° latitudes with increasing SSTs. In addition, the OWZP scheme detects an increase in genesis at 25–40° degrees of latitude, showing a shift in the genesis towards the poles with increasing temperatures (Figure 4). These results agree with the earlier study by Merlis et al. (2016) [33], using uniform SSTs, who showed a decrease in frequency and a shift in the genesis locations towards the poles with increasing SSTs. At higher latitude bands (40–60° and 60–90°), both tracking schemes show a decrease in the TC frequency with increasing temperature (Figure 4).

Figure 3 shows that there is a high density of TCs at polar latitudes. Similar results were found by Chavas and Reed (2019) [17]. In explaining this result, Chavas and Reed (2019) [17] hypothesized that the polar cap region was analogous to an f-plane region, where cyclones dominate. Previous work (e.g., Emanuel 1986 [12]; Chavas and Emanuel 2014 [28]) found that tropical cyclone size scaled approximately with the ratio of potential intensity to the Coriolis parameter. The simulations show higher values of potential intensity for higher SSTs, so in a polar region densely packed with TCs, larger storms might also be fewer, which could be a potential explanation for the results in Figures 3 and 4. Chavas and Reed (2019) [17] found a maximum formation rate at about 30 degrees latitude for their aqua-planet, uniform insolation experiments. We also find a local maximum at 30 degrees of latitude using the OWZP detection scheme (Figure 3), but poleward of this, the genesis density increases. One significant difference in our simulations from those of Chavas and Reed (2019) [17] is that we do not have uniform insolation but keep insolation at terrestrial values. This may be assisting formation at higher latitudes in our results compared with those of Chavas and Reed (2019) [17].

The OWZP and CSIRO scheme-detected TCs have similar maximum lifetimes with a higher number of more prolonged duration storms in the OWZP scheme. The CSIRO TC tracks are a subset of the OWZP tracks in all the three constant SST aqua-planet simulations, similar to the results of the current terrestrial climate simulation (Raavi and Walsh, 2020a) [3].

The fact that they largely share part of their tracks suggests that the response of TC numbers to climate forcing should be similar in the two schemes. A possible exception is at latitudes 30–35°: while both schemes show a local maximum of formation at these latitudes (Figures 3 and 4), formation is considerably higher in the OWZP scheme. Examination of Figure S5 suggests that the OWZP scheme, being a more sensitive detector of weak systems that the CSIRO scheme, detects a substantial number of short-lived disturbances at these latitudes. This may account for the different response to increased SST of TC formation at these latitudes in the OWZP scheme (Figure 4). The TC tracks move poleward due to the beta-drift mechanism. The TCs in the higher latitudes are noted to have longer lifetimes compared to those in the lower latitudes, perhaps due to stronger vorticity and higher upward mass flux, enabling the TCs to survive longer. In general, long-lived TCs appear to be a feature of tropical SST aqua-planet experiments (e.g., Chavas and Reed 2019) [17].

We use storm-centered instantaneous values of large-scale environmental variables to understand the changes in the TC frequency across different latitudes. Of all the variables, we observe that the saturation deficit and stability parameters have the most significant differences between different SST simulations. There is an increase in the saturation deficit and stability of the atmosphere with increasing temperatures. A consistent result was seen in the recent study of Walsh et al. (2020) [10] using the climatological averages of the large-scale variables in similar model experiments. Additionally, there is an increase in stability with latitude in all three constant SST simulations. One possible reason is that increasing precipitation towards the poles heats the upper troposphere at a higher rate, making the atmosphere more stable. Therefore, the atmospheric column is more strongly convectively heated at the poles, thereby increasing the stability compared to the tropics.

One of the hypotheses that explains the reduced TC frequency is the upward mass flux hypothesis [69,70,72]. The reduction in the upward mass flux with increasing surface temperature is found to be a robust result in climate models (Vecchi and Soden, 2007 [73]; Held and Soden, 2006 [74]). The decrease in upward mass flux in the mid-troposphere of the tropics with increasing temperatures is due to the increased stability of the atmosphere (Sugi et al., 2002, 2012 [70]). In the current study, in the lower latitude bands 0–25° and 25–40°, we observe the detected individual storms in the SST25 experiment have more upward mass flux compared to the SST30 experiment. That means there is a reduced upward mass flux with increasing temperatures in the lower latitudes in the aqua-planet simulations. In contrast, at higher latitudes, the storms in the SST30 experiment have higher upward mass flux. Similar trends are observed for the OW parameter and the 10 m wind speed variable, with higher values for SST25 storms at lower latitudes and enhanced values for SST30 at higher latitudes.

Another hypothesis explaining the reduction in the TC frequency is the saturation deficit hypothesis [75,76]. These studies note that increased values of low-level saturation deficit led to reduced TC formations. An idealized study by Rappin et al. (2010) [77] noted that larger values of saturation deficit led to the development of a weaker vortex at a slower rate, thereby leading to reduced cyclone genesis. In the current study, we observe for increasing temperatures that there is an increase in the saturation deficit across all the latitude bands. In contrast, the RH shows prominent differences between different experiments at the lower latitudes, in that there is a decrease in the humidity with increasing temperature. Additionally, Sugi et al. (2015) [78] noted that both these hypotheses explaining the reduction in the TC frequency may be not independent, as with increased saturation deficit the atmosphere is drier and leads to reduced upward mass flux and thereby reduces the potential for TC formation from an initial vortex. In the current study, for all the latitude bands across the globe, we observe a stronger relationship to the stability and the saturation deficit of the atmosphere compared to the reduction in the low-level relative humidity or the reduced upward mass flux. This weaker relationship with omega may be perhaps due to complex interactions between different variables, but we have not examined this issue here.

To further understand the co-influences of the different environmental variables on TC formation with increasing temperatures, we have performed a binary classification for the storms in the SST25 and SST30 experiments. The decision rules obtained from the C4.5 decision tree classification algorithm for the SST25 and SST30 storms (Table 2) show that 700 hPa relative humidity acts as a vital discriminator for the storms at the lower latitudes (0–25° and 25–40°), whereas the OW (which correlates with omega) and WIND variables act as the most significant influencing variables at higher latitudes. Therefore, RH, WIND, and omega or the OW parameter act as the most significant influencing variables for TC formation in the SST25 and SST30 experiments. Moreover, TC formation at the higher latitudes, where there is stronger stability and increased saturation deficits of the atmosphere, requires higher values of the OW or higher values of upward mass flux at the time of genesis. Vortices with lower values of OW or lower values of upward mass flux, thus with a weaker circulation, do not develop under strong, stable conditions. These unfavorable environmental conditions may therefore lead to a reduction in the frequency at those latitudes as not all initial vortices may have the required higher values of the upward mass flux or OW.

With increasing SSTs, we observe a decrease in the number of non-developing tropical depressions, similar to the decrease in developing cases. With increasing temperatures, the reduction in the overall formation rate may be due to a higher saturation deficit and reduced upward mass flux or the non-development of the weaker vortices under more stable atmospheric conditions. From the environmental differences, we observe that the OW parameter and omega act as the most significant discriminating variables for developing versus non-developing depressions under strongly stable and increased saturation deficit atmospheric conditions.

5. Conclusions

In this study, we have shown the response of the global tropical cyclone (TC) and tropical depression (TD) frequency to increased sea surface temperatures (SST) and the associated environmental conditions. For this, we use aqua-planet simulations using a global high-resolution climate model (ACCESS) with three different values of uniform SSTs and two fundamentally different TC detection and tracking schemes. A new pathway for TC formation, the 'marsupial pouch theory', explains the formation of an initial vortex in a closed pouch region within large-scale disturbances. We use the OWZP scheme (a phenomenon-based, resolution-independent method) to detect the locations within the pouches favorable for TC formation. Another TC detection scheme, the CSIRO scheme (a traditional, resolution-dependent method), detects the vortices that resemble observed TCs using wind speed and vorticity thresholds. In the current study, we have found the following results:

(i) Globally, we find an increase in the saturation deficit and stability of the atmosphere with increasing temperatures (Figure 7) and a decrease in the TC frequency using both traditional and phenomenon-based TC tracking schemes (Table 1).

(ii) Although the TCs within the ACCESS model have weaker intensity than in observations, we observe an increase in the frequency of the intense storms with increasing SSTs using both tracking schemes (Figure 1).

In subsequent analysis with the phenomenon-based OWZP scheme, we find the following:

(iii) We observe a reduction in the global mean detections of both developing (TCs) and non-developing tropical depressions with increasing SSTs (Table 1), suggesting that there may be a reduction in the frequency of the weaker storms in a future climate.

(iv) At higher latitudes, which have higher stability compared to the lower latitudes in the SST30 experiment (Figure 7), we notice that the individual TC circulations have higher values of OW and lower values of omega (Figure 8). This result suggests that under more stable atmospheric conditions with increased values of saturation deficit, those initial vortices that have higher strength are more likely to develop into TCs compared to low-strength vortices.

(v) In the OWZP scheme, the TC formation decreases a little in the 0–25° latitude band (Figure 4a), accompanied by lower values of low-level RH, higher omega (i.e., lower values of upward mass flux), higher saturation deficit, and higher atmospheric stability with increasing temperatures (Figures 7 and 8). This result is consistent with earlier hypotheses relating TC formation to reduced upward mass flux and increased saturation deficit. This indicates that in the lower latitudes under increased SSTs, the increased stability of the atmosphere with increasing saturation deficit and reduced upward mass flux led to a decrease in the TC frequency.

(vi) The OWZP scheme identifies a poleward shift in the genesis locations in the 25–40° latitude band with increasing SSTs (Figure 4) due to lower values of vertical wind shear (Figure 8).

(vii) The low-level OW parameter and 500 hPa omega act as significant influencing variables leading to the development of an initial vortex to a TC. The non-developing cases have significantly lower values of OW and upward mass flux (increased omega) compared to the developing cases (Figure 11).

(viii) Overall, the latitudinal variations in the large-scale environmental conditions account for the latitudinal differences in the TC frequency in the OWZP scheme.

From this study, we observe a reduction in tropical depressions (i.e., weaker vortices) and an increase in the number of intense storms with increasing temperatures. Under highly stable atmospheric conditions with higher saturation deficits, only sufficiently strong vortices at the time of genesis develop into TCs. At lower latitudes (0–45°), we observe differences in the TC frequency between the phenomenon-based (OWZP) and traditional (CSIRO) TC tracking schemes, indicating the sensitivity of TC frequency to the formulation of tracking schemes. It is also important to know the influence of vertical wind shear on TC formation and its development with varying meridional SST gradients due to the expansion of the tropics in future climates (Held & Hou, 1980 [79]; Lu et al., 2007 [80]; Staten et al., 2018 [81]). Future studies will focus on other aqua-planet experiments with different meridional SST gradients that lead to specific changes in TC frequency due to the changes in the Hadley circulation and the development of vertical wind shear due to an altered thermal wind balance. Another possible mechanism is that increased values of shear may influence the formation of the shear sheath, a protective layer surrounding a vortex that restricts the inflow of dry air or anticyclonic vorticity supporting the vortex for formation and development (Rutherford et al., 2015 [82], 2018 [83]). In addition, it is essential to compare the current results with simulations that have constant insolation and constant radiative forcing (Shi & Bretherton, 2014 [34]; Merlis et al., 2016 [33]; Chavas & Reed, 2019 [17]), instead of seasonally varying values, in similar uniform SST experiments.

Supplementary Materials: The following are available online at https://www.mdpi.com/article/10.3390/oceans2040045/s1, Figure S1: Diagram showing the tracks of developing and non-developing tropical depressions with F(False) indicating that the core thresholds are not satisfied and T(True) where the thresholds are satisfied. The 3rd consecutive true is the TD decla-ration time for both developing and non-developing TDs. Path A shows that all TSs were once TDs, and path B is a TD that did not develop into TS. Figure S2: Annual mean mass streamfunction (kg s^{-1}) for all experiments, along with differences in values. Figure S3: Mean temperature and zonal wind variation with height, June to August average, for all three experiments. Figure S4: The same as Fig. S3 but for December to February. Figure S5: Tracks of the 280 storms detected using the OWZP scheme (red lines; genesis locations at the green dots) and 214 storms detected by the CSIRO scheme (blue lines; genesis locations given by magenta dots), for August 1990 of the SST25 experiment. Table S1: The initial and core thresholds for tracking TCs using OWZP detection scheme.

Author Contributions: K.J.E.W. supervised the research and provided editorial assistance (20%). P.H.R. carried out the rest of the work (80%). All authors have read and agreed to the published version of the manuscript.

Funding: The authors would like to acknowledge funding from the Australian Research Council under Discovery Project DP15012272. Author Raavi acknowledges the Melbourne India Postgraduate Program for providing her with funding to carry out her PhD at the University of Melbourne. She also thanks the University of Melbourne, Faculty of Science for providing support through the Postgraduate Writing-up Award. We also thank the Australian Research Council Centre of Excellence for Climate Extremes (grant CE170100023) for providing partial funding. We want to thank the ACCESS modeling group and the National Computational Infrastructure system, supported by the Australian government, for providing the model and computing resources on the NCI supercomputer.

Data Availability Statement: The data presented in this study are openly available in University of Melbourne FigShare at doi: 10.26188/c.5721125.

Conflicts of Interest: There are no known conflicts of interest.

Appendix A

C4.5 Classification Algorithm

A classification method is a supervised learning algorithm to separate binary class data into respective classes from the existing datasets based on the background environmental conditions as predictors (Quinlan, 1987 [63], 1993 [84]). Here, we use the C4.5 algorithm to classify the data into the developing and non-developing cases by building a tree-like structure based on thresholds of important environmental variables using a specific measure called "information gain". The information gain is a probabilistic measure that decreases the uncertainty in the dataset after it splits into either a developing or a non-developing class using environmental variables as predictors. This algorithm performs a test at every step and selects an environmental variable that produces an effective separation between both the classes. The environmental variable that has maximum information gain (i.e., best at separating the two classes) is selected as the root of the decision tree. The subsequent splitting of a decision tree stops when a specified minimum leaf size threshold is reached. The minimum leaf size represents the size of the decision tree; a smaller tree is desirable, as it better generalizes the data. A decision tree with overfitting does not perform better for a new test dataset, as it is not generalized well for the overall training dataset. A proper strategy to avoid the overfitting of training samples is to use "pruning", which reduces the complexity of the tree structure by cutting the branches that poorly perform classification. This provides better predictability (Quinlan, 1987 [63]; Hastie et al., 2001 [85]). The current study uses "reduced error pruning" to avoid overfitting and provide a robust classification model. The Weka 3.6.2 open-source software package is used to perform the C4.5 algorithm (https://www.cs.waikato.ac.nz/ml/weka/, accessed on 15 March 2020). We also perform a 10-fold cross-validation to validate the learned decision tree model from training samples to a new test dataset. An optimum value for the "minimum leaf size", the one that gives the highest classification accuracy, is obtained through cross-validation with the reduced error pruning option.

References

1. Bell, R.; Strachan, J.; Vidale, P.L.; Hodges, K.; Roberts, M. Response of Tropical Cyclones to Idealized Climate Change Experiments in a Global High-Resolution Coupled General Circulation Model. *J. Clim.* **2013**, *26*, 7966–7980. [CrossRef]
2. Camargo, S.J.; Wing, A.A. Tropical cyclones in climate models. *Wiley Interdiscip. Rev. Clim. Chang.* **2016**, *7*, 211–237. [CrossRef]
3. Raavi, P.H.; Walsh, K. Sensitivity of Tropical Cyclone Formation to Resolution-Dependent and Independent Tracking Schemes in High-Resolution Climate Model Simulations. *Earth Space Sci.* **2020**, *7*, e2019EA000906. [CrossRef]
4. Shaevitz, D.A.; Camargo, S.J.; Sobel, A.H.; Jonas, J.A.; Kim, D.; Kumar, A.; LaRow, T.E.; Lim, Y.K.; Murakami, H.; Reed, K.A.; et al. Characteristics of tropical cyclones in high-resolution models in the present climate. *J. Adv. Model. Earth Syst.* **2014**, *6*, 1154–1172. [CrossRef]
5. Sharmila, S.; Walsh, K.J.E.; Thatcher, M.; Wales, S.; Utembe, S. Real world and tropical cyclone world. Part I: High-resolution climate model verification. *J. Clim.* **2020**, *33*, 1455–1472. [CrossRef]
6. Strachan, J.; Vidale, P.L.; Hodges, K.I.; Roberts, M.; Demory, M.-E. Investigating Global Tropical Cyclone Activity with a Hierarchy of AGCMs: The Role of Model Resolution. *J. Clim.* **2013**, *26*, 133–152. [CrossRef]
7. Tory, K.J.; Chand, S.S.; Dare, R.A.; McBride, J.L. An Assessment of a Model-, Grid-, and Basin-Independent Tropical Cyclone Detection Scheme in Selected CMIP3 Global Climate Models. *J. Clim.* **2013**, *26*, 5508–5522. [CrossRef]
8. Tory, K.J.; Chand, S.S.; McBride, J.L.; Ye, H.; Dare, R.A. Projected changes in late-twenty-first-century tropical cyclone frequency in 13 coupled climate models from phase 5 of the Coupled Model Intercomparison Project. *J. Clim.* **2013**, *26*, 9946–9959. [CrossRef]
9. Walsh, K.J.; Camargo, S.J.; Vecchi, G.A.; Daloz, A.S.; Elsner, J.; Emanuel, K.; Horn, M.; Lim, Y.K.; Roberts, M.; Patricola, C.; et al. Hurricanes and climate: The US CLIVAR working group on hurricanes. *Bull. Am. Meteorol. Soc.* **2015**, *96*, 997–1017. [CrossRef]
10. Walsh, K.J.E.; Sharmila, S.; Thatcher, M.; Wales, S.; Utembe, S.; Vaughan, A. Real World and Tropical Cyclone World. Part II: Sensitivity of Tropical Cyclone Formation to Uniform and Meridionally Varying Sea Surface Temperatures under Aquaplanet Conditions. *J. Clim.* **2020**, *33*, 1473–1486. [CrossRef]
11. Zhao, M.; Held, I.M.; Lin, S.-J.; Vecchi, G. Simulations of Global Hurricane Climatology, Interannual Variability, and Response to Global Warming Using a 50-km Resolution GCM. *J. Clim.* **2009**, *22*, 6653–6678. [CrossRef]
12. Emanuel, K.A. An air-sea interaction theory for tropical cyclones. Part I: Steady-state maintenance. *J. Atmos. Sci.* **1986**, *43*, 585–605. [CrossRef]
13. Emanuel, K.A. The Maximum Intensity of Hurricanes. *J. Atmos. Sci.* **1988**, *45*, 1143–1155. [CrossRef]
14. Knutson, T.; Camargo, S.; Chan, J.C.L.; Emanuel, K.; Ho, C.-H.; Kossin, J.; Mohapatra, M.; Satoh, M.; Sugi, M.; Walsh, K.; et al. Tropical Cyclones and Climate Change Assessment: Part II: Projected Response to Anthropogenic Warming. *Bull. Am. Meteorol. Soc.* **2020**, *101*, E303–E322. [CrossRef]
15. Camargo, S.J.; Tippett, M.K.; Sobel, A.H.; Vecchi, G.A.; Zhao, M. Testing the performance of tropical cyclone genesis indices in future climates using the HiRAM model. *J. Clim.* **2014**, *27*, 9171–9196. [CrossRef]
16. Walsh, K.J.; McBride, J.L.; Klotzbach, P.J.; Balachandran, S.; Camargo, S.J.; Holland, G.; Knutson, T.R.; Kossin, J.P.; Lee, T.C.; Sobel, A.; et al. Tropical cyclones and climate change. *Wiley Interdiscip. Rev. Clim. Chang.* **2016**, *7*, 65–89. [CrossRef]
17. Chavas, D.R.; Reed, K.A. Dynamical Aquaplanet Experiments with Uniform Thermal Forcing: System Dynamics and Implications for Tropical Cyclone Genesis and Size. *J. Atmos. Sci.* **2019**, *76*, 2257–2274. [CrossRef]
18. Vu, T.; Kieu, C.; Chavas, D.; Wang, Q. A Numerical Study of the Global Formation of Tropical Cyclones. *J. Adv. Model. Earth Syst.* **2021**, *13*, e2020MS002207. [CrossRef]
19. Davis, C.A. The formation of moist vortices and tropical cyclones in idealized simulations. *J. Atmos. Sci.* **2015**, *72*, 3499–3516. [CrossRef]
20. Held, I.M.; Zhao, M. Horizontally Homogeneous Rotating Radiative–Convective Equilibria at GCM Resolution. *J. Atmos. Sci.* **2008**, *65*, 2003–2013. [CrossRef]
21. Khairoutdinov, M.; Emanuel, K. Rotating radiative-convective equilibrium simulated by a cloud-resolving model. *J. Adv. Model. Earth Syst.* **2013**, *5*, 816–825. [CrossRef]
22. Knutson, T.R.; Tuleya, R.E. Impact of CO_2-induced warming on simulated hurricane intensity and precipitation: Sensitivity to the choice of climate model and convective parameterization. *J. Clim.* **2004**, *17*, 3477–3495. [CrossRef]
23. Murthy, V.S.; Boos, W.R. Role of Surface Enthalpy Fluxes in Idealized Simulations of Tropical Depression Spinup. *J. Atmos. Sci.* **2018**, *75*, 1811–1831. [CrossRef]
24. Nolan, D.S.; Rappin, E.D.; Emanuel, K.A. Tropical cyclogenesis sensitivity to environmental parameters in radiative-convective equilibrium. *Q. J. R. Meteorol. Soc.* **2007**, *133*, 2085–2107. [CrossRef]
25. Nolan, D.S.; Rappin, E.D. Increased sensitivity of tropical cyclogenesis to wind shear in higher SST environments. *Geophys. Res. Lett.* **2008**, *35*, 1–7. [CrossRef]
26. Wing, A.A.; Camargo, S.; Sobel, A. Role of Radiative–Convective Feedbacks in Spontaneous Tropical Cyclogenesis in Idealized Numerical Simulations. *J. Atmos. Sci.* **2016**, *73*, 2633–2642. [CrossRef]
27. Zhou, W.; Held, I.M.; Garner, S.T. Parameter Study of Tropical Cyclones in Rotating Radiative—Convective Equilibrium with Column Physics and Resolution of a 25-km GCM. *J. Atmos. Sci.* **2014**, *71*, 1058–1069. [CrossRef]

28. Chavas, D.R.; Emanuel, K. Equilibrium Tropical Cyclone Size in an Idealized State of Axisymmetric Radiative–Convective Equilibrium. *J. Atmos. Sci.* **2014**, *71*, 1663–1680. [CrossRef]
29. Hayashi, Y.-Y.; Sumi, A. The 30–40 day oscillations simulated in an "aqua planet" model. *J. Meteorol. Soc. Jpn.* **1986**, *64*, 451–467. [CrossRef]
30. Merlis, T.M.; Held, I.M. Aquaplanet Simulations of Tropical Cyclones. *Curr. Clim. Chang. Rep.* **2019**, *5*, 185–195. [CrossRef]
31. Merlis, T.M.; Zhao, M.; Held, I.M. The sensitivity of hurricane frequency to ITCZ changes and radiatively forced warming in aquaplanet simulations. *Geophys. Res. Lett.* **2013**, *40*, 4109–4114. [CrossRef]
32. Ballinger, A.; Merlis, T.M.; Held, I.M.; Zhao, M. The Sensitivity of Tropical Cyclone Activity to Off-Equatorial Thermal Forcing in Aquaplanet Simulations. *J. Atmos. Sci.* **2015**, *72*, 2286–2302. [CrossRef]
33. Merlis, T.M.; Zhou, W.; Held, I.M.; Zhao, M. Surface temperature dependence of tropical cyclone-permitting simulations in a spherical model with uniform thermal forcing. *Geophys. Res. Lett.* **2016**, *43*, 2859–2865. [CrossRef]
34. Shi, X.; Bretherton, C.S. Large-scale character of an atmosphere in rotating radiative-convective equilibrium. *J. Adv. Model. Earth Syst.* **2014**, *6*, 616–629. [CrossRef]
35. Reed, K.A.; Chavas, D.R. Uniformly rotating global radiative-convective equilibrium in the Community Atmosphere Model, version 5. *J. Adv. Model. Earth Syst.* **2015**, *7*, 1938–1955. [CrossRef]
36. Dunkerton, T.J.; Montgomery, M.T.; Wang, Z. Tropical cyclogenesis in a tropical wave critical layer: Easterly waves. *Atmos. Chem. Phys.* **2009**, *9*, 5587–5646. [CrossRef]
37. Wang, Z.; Montgomery, M.T.; Fritz, C. A First Look at the Structure of the Wave Pouch during the 2009 PREDICT–GRIP Dry Runs over the Atlantic. *Mon. Weather. Rev.* **2012**, *140*, 1144–1163. [CrossRef]
38. Tory, K.J.; Dare, R.A.; Davidson, N.E.; McBride, J.L.; Chand, S.S. The importance of low-deformation vorticity in tropical cyclone formation. *Atmos. Chem. Phys.* **2013**, *13*, 2115–2132. [CrossRef]
39. Tory, K.J.; Ye, H.; Dare, R.A. Understanding the geographic distribution of tropical cyclone formation for applications in climate models. *Clim. Dyn.* **2018**, *50*, 2489–2512. [CrossRef]
40. Kirtman, B.P.; Schneider, E.K. A spontaneously generated tropical atmospheric general circulation. *J. Atmos. Sci.* **2000**, *57*, 2080–2093. [CrossRef]
41. Ullrich, P.A.; Zarzycki, C.M. TempestExtremes: A framework for scale-insensitive pointwise feature tracking on unstructured grids. *Geosci. Model Dev.* **2017**, *10*, 1069–1090. [CrossRef]
42. Emanuel, K.A. Downscaling CMIP5 climate models shows increased tropical cyclone activity over the 21st century. *Proc. Natl. Acad. Sci. USA* **2013**, *110*, 12219–12224. [CrossRef]
43. Bhatia, K.; Vecchi, G.; Murakami, H.; Underwood, S.; Kossin, J. Projected Response of Tropical Cyclone Intensity and Intensification in a Global Climate Model. *J. Clim.* **2018**, *31*, 8281–8303. [CrossRef]
44. Vecchi, G.A.; Delworth, T.L.; Murakami, H.; Underwood, S.D.; Wittenberg, A.T.; Zeng, F.; Zhang, W.; Baldwin, J.W.; Bhatia, K.T.; Cooke, W.; et al. Tropical cyclone sensitivities to CO_2 doubling: Roles of atmospheric resolution, synoptic variability and background climate changes. *Clim. Dyn.* **2019**, *53*, 5999–6033. [CrossRef]
45. Sugi, M.; Yamada, Y.; Yoshida, K.; Mizuta, R.; Nakano, M.; Kodama, C.; Satoh, M. Future Changes in the Global Frequency of Tropical Cyclone Seeds. *SOLA* **2020**, *16*, 70–74. [CrossRef]
46. Yamada, Y.; Kodama, C.; Satoh, M.; Sugi, M.; Roberts, M.J.; Mizuta, R.; Noda, A.T.; Nasuno, T.; Nakano, M.; Vidale, P.L. Evaluation of the contribution of tropical cyclone seeds to changes in tropical cyclone frequency due to global warming in high-resolution multi-model ensemble simulations. *Prog. Earth Planet. Sci.* **2021**, *8*, 11. [CrossRef]
47. Hsieh, T.-L.; Vecchi, G.A.; Yang, W.; Held, I.M.; Garner, S.T. Large-scale control on the frequency of tropical cyclones and seeds: A consistent relationship across a hierarchy of global atmospheric models. *Clim. Dyn.* **2020**, *55*, 3177–3196. [CrossRef]
48. Tang, B.; Emanuel, K. A Ventilation Index for Tropical Cyclones. *Bull. Am. Meteorol. Soc.* **2012**, *93*, 1901–1912. [CrossRef]
49. Horn, M.; Walsh, K.; Zhao, M.; Camargo, S.J.; Scoccimarro, E.; Murakami, H.; Wang, H.; Ballinger, A.; Kumar, A.; Shaevitz, D.A.; et al. Tracking scheme dependence of simulated tropical cyclone response to idealized climate simulations. *J. Clim.* **2014**, *27*, 9197–9213. [CrossRef]
50. Bi, D.; Dix, M.; Marsland, S.; O'Farrell, S.; Rashid, H.; Uotila, P.; Hirst, A.; Kowalczyk, E.; Golebiewski, M.; Sullivan, A.; et al. The ACCESS coupled model: Description, control climate and evaluation. *Aust. Meteorol. Oceanogr. J.* **2013**, *63*, 41–64. [CrossRef]
51. Best, M.J.; Pryor, M.; Clark, D.B.; Rooney, G.G.; Essery, R.; Ménard, C.B.; Edwards, J.M.; Hendry, M.A.; Porson, A.; Gedney, N.; et al. The Joint UK Land Environment Simulator (JULES), model description—Part 1: Energy and water fluxes. *Geosci. Model Dev.* **2011**, *4*, 677–699. [CrossRef]
52. Walters, D.; Brooks, M.; Boutle, I.; Melvin, T.; Stratton, R.; Vosper, S.; Wells, H.; Williams, K.; Wood, N.; Allen, T.; et al. The Met Office unified model global atmosphere 6.0/6.1 and JULES global land 6.0/6.1 configurations. *Geosci. Model Dev.* **2017**, *10*, 1487–1520. [CrossRef]
53. Wilson, D.R.; Bushell, A.C.; Kerr-Munslow, A.M.; Price, J.D.; Morcrette, C.J. PC2: A prognostic cloud fraction and condensation scheme. I: Scheme description. *Q. J. R. Meteorol. Soc.* **2008**, *134*, 2093–2107. [CrossRef]
54. Edwards, J.M.; Slingo, A. Studies with a flexible new radiation code. I: Choosing a configuration for a large-scale model. *Q. J. R. Meteorol. Soc.* **1996**, *122*, 689–719. [CrossRef]
55. Lock, A.P.; Brown, A.R.; Bush, M.R.; Martin, G.M.; Smith, R.N.B. A new boundary layer mixing scheme. Part I: Scheme description and single-column model tests. *Mon. Weather Rev.* **2000**, *128*, 3187–3199. [CrossRef]

56. Lock, A.P. The Numerical Representation of Entrainment in Parameterizations of Boundary Layer Turbulent Mixing. *Mon. Weather Rev.* **2001**, *129*, 1148–1163. [CrossRef]
57. Brown, A.R.; Beare, R.J.; Edwards, J.M.; Lock, A.P.; Keogh, S.J.; Milton, S.F.; Walters, D. Upgrades to the Boundary-Layer Scheme in the Met Office Numerical Weather Prediction Model. *Bound.-Layer Meteorol.* **2008**, *128*, 117–132. [CrossRef]
58. Kanamitsu, M.; Ebisuzaki, W.; Woollen, J.; Yang, S.-K.; Hnilo, J.J.; Fiorino, M.; Potter, G.L. NCEP–DOE AMIP-II Reanalysis (R-2). *Bull. Am. Meteorol. Soc.* **2002**, *83*, 1631–1643. [CrossRef]
59. Walsh, K.J.E.; Fiorino, M.; Landsea, C.W.; McInnes, K.L. Objectively Determined Resolution-Dependent Threshold Criteria for the Detection of Tropical Cyclones in Climate Models and Reanalyses. *J. Clim.* **2007**, *20*, 2307–2314. [CrossRef]
60. Tory, K.J.; Chand, S.S.; Dare, R.A.; McBride, J.L. The Development and Assessment of a Model-, Grid-, and Basin-Independent Tropical Cyclone Detection Scheme. *J. Clim.* **2013**, *26*, 5493–5507. [CrossRef]
61. Bell, S.S.; Chand, S.S.; Tory, K.J.; Turville, C. Statistical Assessment of the OWZ Tropical Cyclone Tracking Scheme in ERA-Interim. *J. Clim.* **2018**, *31*, 2217–2232. [CrossRef]
62. Raavi, P.H.; Walsh, K.J.E. Basinwise statistical analysis of factors limiting tropical storm formation from an initial tropical circulation. *J. Geophys. Res. Atmos.* **2020**, *125*, e2019JD032006. [CrossRef]
63. Quinlan, J. Decision trees as probabilistic classifiers. In Proceedings of the Fourth International Workshop on Machine Learning, Irvine, CA, USA, 22–25 June 1987; pp. 31–37.
64. Sharmila, S.; Walsh, K.J.E. Recent poleward shift of tropical cyclone formation and its link to tropical expansion. *Nat. Clim. Chang.* **2018**, *8*, 730–736. [CrossRef]
65. Peixoto, J.P.; Oort, A.H. *Physics of Climate*; American Institute of Physics: College Park, MD, USA, 1992; 520p.
66. Holland, G.J. Tropical Cyclone Motion: Environmental Interaction Plus a Beta Effect. *J. Atmos. Sci.* **1983**, *40*, 328–342. [CrossRef]
67. Smith, R.K. An analytic theory of tropical-cyclone motion in a barotropic shear flow. *Q. J. R. Meteorol. Soc.* **1991**, *117*, 685–714. [CrossRef]
68. Wang, Y.; Holland, G.J. The Beta Drift of Baroclinic Vortices. Part I: Adiabatic Vortices. *J. Atmos. Sci.* **1996**, *53*, 411–427. [CrossRef]
69. Sugi, M.; Noda, A.; Sato, N. Influence of the Global Warming on Tropical Cyclone Climatology: An Experiment with the JMA Global Model. *J. Meteorol. Soc. Jpn.* **2002**, *80*, 249–272. [CrossRef]
70. Sugi, M.; Murakami, H.; Yoshimura, J. On the Mechanism of Tropical Cyclone Frequency Changes Due to Global Warming. *J. Meteorol. Soc. Jpn.* **2012**, *90A*, 397–408. [CrossRef]
71. Knutson, T.R.; Sirutis, J.J.; Zhao, M.; Tuleya, R.E.; Bender, M.; Vecchi, G.A.; Villarini, G.; Chavas, D. Global projections of intense tropical cyclone activity for the late twenty-first century from dynamical downscaling of CMIP5/RCP4. 5 scenarios. *J. Clim.* **2015**, *28*, 7203–7224. [CrossRef]
72. Held, I.M.; Zhao, M. The Response of Tropical Cyclone Statistics to an Increase in CO_2 with Fixed Sea Surface Temperatures. *J. Clim.* **2011**, *24*, 5353–5364. [CrossRef]
73. Vecchi, G.A.; Soden, B. Global Warming and the Weakening of the Tropical Circulation. *J. Clim.* **2007**, *20*, 4316–4340. [CrossRef]
74. Held, I.M.; Soden, B.J. Robust Responses of the Hydrological Cycle to Global Warming. *J. Clim.* **2006**, *19*, 5686–5699. [CrossRef]
75. Emanuel, K.; Sundararajan, R.; Williams, J. Hurricanes and Global Warming: Results from Downscaling IPCC AR4 Simulations. *Bull. Am. Meteorol. Soc.* **2008**, *89*, 347–368. [CrossRef]
76. Emanuel, K. Tropical cyclone activity downscaled from NOAA-CIRES Reanalysis, 1908–1958. *J. Adv. Model. Earth Syst.* **2010**, *2*, 1–12. [CrossRef]
77. Rappin, E.D.; Nolan, D.S.; Emanuel, K.A. Thermodynamic control of tropical cyclogenesis in environments of radiative-convective equilibrium with shear. *Q. J. R. Meteorol. Soc.* **2010**, *136*, 1954–1971. [CrossRef]
78. Sugi, M.; Yoshida, K.; Murakami, H. More tropical cyclones in a cooler climate? *Geophys. Res. Lett.* **2015**, *42*, 6780–6784. [CrossRef]
79. Held, I.M.; Hou, A.Y. Nonlinear axially symmetric circulations in a nearly inviscid atmosphere. *J. Atmos. Sci.* **1980**, *37*, 515–533. [CrossRef]
80. Lu, J.; Vecchi, G.; Reichler, T. Expansion of the Hadley cell under global warming. *Geophys. Res. Lett.* **2007**, *34*, 1–5. [CrossRef]
81. Staten, P.; Lu, J.; Grise, K.M.; Davis, S.M.; Birner, T. Re-examining tropical expansion. *Nat. Clim. Chang.* **2018**, *8*, 768–775. [CrossRef]
82. Rutherford, B.; Dunkerton, T.J.; Montgomery, M.T. Lagrangian vortices in developing tropical cyclones. *Q. J. R. Meteorol. Soc.* **2015**, *141*, 3344–3354. [CrossRef]
83. Rutherford, B.; Boothe, M.A.; Dunkerton, T.J.; Montgomery, M.T. Dynamical properties of developing tropical cyclones using Lagrangian flow topology. *Q. J. R. Meteorol. Soc.* **2018**, *144*, 218–230. [CrossRef]
84. Quinlan, J. *C4.5: Programs for Machine Learning*; Morgan Kaufmann: Burlington, MA, USA, 1993.
85. Hastie, T.; Tibshirani, R.; Friedman, J.J.H. *The Elements of Statistical Learning: Data Mining, Inference, and Prediction*; Springer Series in Statistics; Springer: Berlin/Heidelberg, Germany, 2001; p. 552.

MDPI

St. Alban-Anlage 66

4052 Basel

Switzerland

Tel. +41 61 683 77 34

Fax +41 61 302 89 18

www.mdpi.com

Oceans Editorial Office

E-mail: oceans@mdpi.com

www.mdpi.com/journal/oceans